Insect Growth Regulator and Rice Moth

The Author

Dr. Shri Krishna Tiwari, Ph. D., Asst. Professor, Department of Zoology, D.D.U. Gorakhpur University, Gorakhpur, U. P., India, has vast research experience of about 31 years, currently engaged in research pertaining to management of stored cereal pests through the application of natural plant products, synthetic pyrethroids and a few juvenile hormone analogues like, fenoxycarb having insecticidal properties.

Insect Growth Regulator and Rice Moth

Dr. S.K. Tiwari

Department of Zoology
D.D.U. Gorakhpur University
Gorakhpur – 273 009
Uttar Pradesh

BIOTECH

2019

BIOTECH BOOKS®

ISBN 978-81-7622-439-0

Published by: **BIOTECH BOOKS®**
4762-63/23, Ansari Road, Darya Ganj,
New Delhi - 110 002
Phone: 09899295259
E-mail: biotechbooks@yahoo.co.in

Typeset at: **Classic Computer Services**
Delhi - 110 035

Printed at: **Replika Press Pvt. Ltd.**

PRINTED IN INDIA

Acknowledgement

I am extremely indebted to Prof. Rajendra Singh, Former Head, Department of Zoology, D.D.U. University of Gorakhpur, Gorakhpur, U.P., India for providing necessary laboratory facilities, valuable guidance, constant inspiration and encouragement throughout this research work.

I am cordially grateful to Prof. Ajay Kumar Srivastav, Prof. V. B. Upadhyaya, Prof. D. K. Singh, Prof. B.A. Ansari, Prof. Ajay Singh, Prof. V.B. Kushwaha, Prof. Ravikant Upadhyay and Prof. P.H. Pathak and all the teachers of this department for providing their constant moral support and encouragement.

I am sincerely thankful to my research student Dr. Akansha Singh whose continuous effort and dedication throughout the experiments made possible to complete this tedious and rare work. I am also thankful to my research students Dr. Chandra Shekhar Pathak, Dr. Snehlata Shukla, Dr. Priyanka Tripathi, Dr. Awanish Chandra and Dr. Ajeet Kumar Tiwari for their valuable and active co-operation during the entire period of this investigation.

S.K. Tiwari

Preface

Insect pest management is still a puzzling problem. Modern insecticide research started almost 65 years ago with the chlorinated hydrocarbons followed by organophosphates, methylcarbamates, botanicals and synthetic pyrethroids (Mojaver and Bandani, 2010). Persistent application of these conventional organic insecticides affect immune system of insects, develop resistance, cause toxicity to non-target organisms like parasites , predators and beneficial insects, residue in food grains, biomagnifications and ofcourse pollute our own environment due to non-biodegradability. In the search for safer insecticide technologies i.e. more selective mode of action and reduced risks for non-target organisms and the environment, progress has been made in the last two decades with the development of natural and synthetic compounds capable of interfering with the processes of growth, development and metamorphosis of the target insects (Smet *et al.*, 1989, 1991 and Oberlander *et al.*, 1997). Williams (1967) suggested application of Insect Growth Regulators "Third Generation Pesticides" as environmentally safe control agents to which the insect will be unable to develop resistance. The development of highly potent synthetic analogues of JH, which were several folds more active than the natural hormone, gave credence to William's claim (Henric *et al.*, 1973).

I feel pleasure that the book entitled "Insect Growth Regulator and Rice Moth" exploring knowledge pertaining to the role of insect growth regulator against biology and reproductive potential of *C. cephalonica* will be of academic importance for entomology intellectuals. This work will devise ways and means for the effective control of rice moth in particular and lepidopterus pests in general so that evolution of new generation of the pest could be considerably restricted in eco-friendly way without harming the non-target organisms.

The author is thankful to M/s Biotech Books, New Delhi, for their suggestions and keen interest in publishing this book.

Dr. S.K. Tiwari

Contents

List of Figures

List of Tables

Chapter 1

Introduction

According to Grain Depot Fund Prospectus (2011), India is one of the largest grain producers in the world. It produces 200 million MT (metric tons) of wheat and rice annually. According to Nagpal and Kumar (2012) post harvest losses in India amount to 12-16 million metric tons of food grain each year, an amount that the World Bank estimates could feed one-third of India's poor. The monetary value of these losses amounts to more than Rs 50,000 crores per year (Singh, 2010). The post harvest losses of food grains have been estimated to the extent of about 10 per cent of the total productions of which losses during storage alone are estimated to be 6.5 per cent. The insects alone damage 2.5 per cent of the food grains in storage. Seventy five per cent of total grain loss occurs at farm level. Majority of such losses occur by inadequate storage (22 per cent), drying of crops (15 per cent), transportation (12 per cent) and threshing (10 per cent) while rest of the losses occur by scattering among several other activities (Grain Depot Fund Prospectus, 2011). An estimation of United States tells that in developed countries the average minimum overall losses from biological degradation are 10 per cent (National Research Council, 1978) while in developing countries that estimate may be upto 20 per cent. In sub-Saharan Africa losses are estimated to be around $4 billion a year (FAO and World Bank, 2011).

Storage loss of food grains at the level of Government and its agencies such as Food Corporation of India, Central and State Warehousing Corporations and State Civil Supplies Departments/Corporations have been reduced to the minimum. However, 60-70 per cent of the total production are retained by the farmers for their own food, cattle feed, seed *etc.* and they generally store their grains in traditional storage structures where maximum loss occur that require intensive care of pest management.

The presence of insects in stored foods directly affects both the quality and quantity of the commodity (Wilbur and Mills, 1985; Hill, 1990 and Burkholder and Faustini, 1991). High grain temperatures and moisture, along with broken kernels, provide conditions that accelerate insect development within the grain mass, increasing grain losses. Storage infestations may originate in the field by highly mobile insects leaving the storage site and flying to grain standing in the field. They may also move to newly stored grain from fields and infested grain bins nearby. Insect populations can reach high levels when left unchecked in grain bins, subfloors, or aeration ducts, and in grain-moving equipment or discarded grain. Insects damage grain by boring holes into the kernels and reducing grain quality through weight, nutritional, or quality loss, spreading and encouraging mold germination, adding to the fatty acid content of the grain, and leaving quantities of uric acid that cause grain rancidity. While feeding, insects also create fines and broken kernels that reduce airflow through the grain when aeration fans are used. This reduction in air flow can cause an increase in temperature, compounding the problem. In addition to the direct damage, the presence of insects in a grain sample can result in cash discounts for the grain.

In 2040, it is estimated that India will need an incremental 150 million MT of grain just to maintain current levels of nutrition in the population (Grain Depot Fund Prospectus, 2011). Therefore reducing the food losses offers an important way of increasing food availability without requiring additional production resources and in less developed countries it can contribute to rural development and poverty reduction.

Control of insect pests is a puzzling problem since many decades. *Corcyra cephalonica* Stainton, commonly known as rice moth, is a severe pest of stored cereals and cereal products in Asia, Africa, Europe, North America and other tropical and subtropical regions of the world. This moth was first identified and reported by Stainton (1866), who named it *Melissoblaptes cephalonica.* Later, Ragonot (1885) gave it the generic name *Corcyra.* The only recognized species of this genus is *cephalonica.* Ayyar (1919) made the first record of *Corcyra cephalonica.* This moth is believed to be of eastern origin but has become a cosmopolitan species. Its larval stages cause serious damage to rice, gram, sorghum, maize, groundnut, cotton seeds, peanuts, linseeds, raisins, nutmeg, currants, chocolates, army biscuits and milled products (Chittenden, 1919; Ayyar, 1919; Ayyar, 1934; Munro and Thompson, 1929; Richards and Herford, 1930; Noyes, 1930; Herford, 1933; Atwal, 1976 and Piltz, 1977). Its mode of damage, with special reference to economic importance, resembles to that of Mediterranean flour moth, *Ephestia kuehniella* Zell., almond moth, *E. cautella* Walk., cacao moth, *E. elutella* Hbn. and raisin moth, *E. figulilella* Gregs. The other equally striking pests of cereals which resemble *C. cephalonica* are the Indian meal moth, *Plodia interpunctella* Hbn., pink corn worm, *Pyroderces rileyi* Wals and lesser grain borer, *Rhizopertha dominica.* Some other lepidopterous pests that deserve attention in relation to *C. cephalonica* are wolfmoth, *Nemapogon granella* (L.), reported from most parts of the world infesting all kinds of grains both in the field and in storage and meal snout moth, *Pyralis tarinalis* (L.) which particularly attracts to bins containing grain slightly out of conditions.

C. cephalonica individuals of both sexes possess ochre-yellowish green pointed forewings and some what grey white hind wings. Their wing span ranges from 14-24 mm. Females can be easily distinguished on the basis of presence of pronounced snout in their head region and comparatively larger in size than male due to relatively elongated and broad abdomen. The moths are active mostly at night and start mating just after emergence. After successful mating, the female starts egg laying within 24 h of mating and continue to do so for 5-7 days of its life-span. The eggs are laid singly or in groups of 3-5. A single female may lay 250-495 eggs during its life span. The average length and breadth of egg are 0.42 and 0.31mm, respectively (Devi *et al.*, 2013). The eggs hatch within the 3-4 days at normal room temperature varying between 25°C and 27°C into slender, elongated larvae which feed under silken web-like shelters, preferring the partially damaged grains. They develop, followed by 4 moults possessing five instars, into brownish pupae inside silken cocoons (Tiwari, 1987). Jagadish *et al.* (2009) recorded the larval development occur inside the grain cluster, followed by 5 moults possessing six instars. The larval period range from 4 to 5, 5 to 6, 3 to 4, 5 to 7 and 8 to 10 days for first, second, third, fourth, fifth and sixth instar respectively. The total developmental period occupied, 41 to 59 days on foxtail millet, *Setaria italica* (Jagadish *et al.*, 2009); 57 days on sorghum (Ayyar, 1934) and 47 to 57 days on broken maize (Seshagir Rao, 1954). The pupal stage lasts for about 9-14 days after which adults emerge out and live for more than one week without feeding. The adult longevity of virgin male is 9-12 days and mated male is 4-5 days and for female it is 7-9 days (Jagadish *et al.*, 2009). Damage to the stored material, by feeding, is caused by the caterpillars only. The larvae are dull white, with long fine hair and dark brown head. Larvae spin silken threads as they feed and web grains, debris and other particles into galleries in which they live and feed. It is noteworthy that females are able to lay eggs without mating but there are no viability of parthenogenically laid eggs *i.e.* they never hatch (Krishna and Narain, 1976).

Modern insecticide research started almost 65 years ago with the chlorinated hydrocarbons, followed shortly by the organophosphates, methylcarbamates, botanicals and synthetic pyrethroids (Mojaver and Bandani, 2010). The use of these conventional organic insecticides to control insect pests has given rise to problems of the proliferation of resistance and accumulation of residue in the environment with adverse ecological effects (Hoffmann and Lorenz, 1997). In the search for safer insecticide technologies, *i.e.* more selective mode of action and reduced risks for non-target organisms and the environment, progress has been made in the last two decades with the development of natural and synthetic compounds capable of interfering with the processes of growth, development and metamorphosis of the target insects (Smet *et al.*, 1989, 1991 and Oberlander *et al.*, 1997).

Considerable measures have been adopted in relation to application of organic pesticides since World War II, which brought inestimable benefits to humanity in terms of human lives saved, diminished sufferings and economic gain (Smith and Bosch, 1967 and Metcalf, 1968).

Since, the organic pesticides were cheap, effective in small quantities, persistent, broadly toxic against whole complex of pests, and require little labour for their deployment, it is no wonder that modern pesticides displaced the biological, environmental and cultural controls which integrated together and sometimes in combination with the earlier inorganic, botanical and raw organic chemical pesticides, formed the prewar pest control arsenal. The new organic synthetic pesticides could be used alone with resulting high levels of pest mortality. Hence, they eliminated the need for many of the earlier pest control schemes that employed a combination of suppressive measures such as ploughing, crop rotation, selective burning and are less labour intensive than these integrated schemes. The successful chemical control has evolved into a highly significant bioactivity. In 1971, $3.4 billion worth of chemical pesticides was applied on a worldwide basis for agricultural (including forestry), industrial and household use (Anonymous, 1973).

Ordinarily, the control measures in stores are based on fumigation with chemicals like hydrogen phosphate. Residues and insect resistance are reasons for potentially limiting the use of fumigation with chemicals in the near future (World Metrological Organisation, 1994).

In recent years, there has been great concern over the toxicity of pesticides on non-target organisms and the environment. Although, chemical pesticides are invaluable in controlling insect populations both in the field and storage, their indiscriminate use has resulted in the destruction of beneficial insects and has caused environmental hazards (Smith and Bosch, 1967; Marx, 1977 and Pimental, 1983). Moreover, insecticide resistance has already developed in many insects which is now a great concern in post-harvest ecosystems throughout the word (Subramanyam and Hagstrum, 1995 and Arthur, 1996). Persistent use of synthetic organic insecticides affect immune system of insects, develop resistance and ofcourse pollute our own environment due to non-biodegradability, biomagnification and toxicity to non-target organisms.

In such condition, there is a need for new alternatives to traditional insecticides used in stored product pest management (Arthur, 1996; Dunkel and Sears, 1998; Mbata and Phillips, 2001 and Arthur and Phillips, 2003). In this regard, the insect growth regulators (IGRs) (Fox, 1990), which mimic insect's hormone and regulate the insect population through the disruption of moulting and metamorphosis (Williams, 1956 and Oberlander *et al.*, 1997) have captured the interest of stored product entomologists. The first use of IGRs against stored product pests was reported by Thomas and Bhatnagar-Thomas (1968) but they have only been tested on a small number of insect species. The term IGR was designed by Staal (1975) to describe a class of bio-rational compounds. Through selectivity of action, these compounds appear to fit the requirements for "Third Generation Pesticides" (Williams, 1967) that disrupt the normal development of several species of insects (Henrick *et al.*, 1973). These compounds are highly effective against various insects attacking stored products and other pests that have become resistant to organic insecticides. Meanwhile, all these compounds are less toxic to mammals and non-target organisms because of their non-toxic effect and their quick disintegrating abilities (Carter, 1975; Staal, 1975; Zurflueh, 1976; Oberlander *et al.*, 1978, 1979; Ishaaya *et al.*, 1987; Ishaaya and Horowtz, 1998; Kostyukovsky *et al.*, 2000 and Parthasarathy *et al.*, 2012).

IGRs generally control insects either through regulation of metamorphosis or interference with reproduction (Riddiford and Truman, 1978). On the basis of the mode of action, workers (Willis, 1974; Staal, 1975; Marx, 1977; Bengston, 1986; Reynolds, 1987 and Wing and Aller, 1990) grouped IGRs into three categories:

(i) Juvenile hormone (JHs) and their analogues (JHAs) also called as juvenoids
(ii) Ecdysone agonists and
(iii) Chitin synthesis inhibitors (CSIs) or moult inhibitors (MIs).

The first category belongs to a group of commercial insecticides that mimic the mode of action of the insect growth and development hormones *i.e.* the sesquiterpenoid juvenile hormones, the second category belongs to a group of insecticide that mimic steroidal 20-hydroxyecdysone and the third class belongs to a group of insecticides that disrupt growth and development by interfering with chitin synthesis in target insect pests (Pener and Dhadialla, 2012). Recently, Pener and Dhadialla (2012) proposed the use of term "Insect Growth Disrupters" instead of "Insect Growth Regulators".

To understand the action of insect hormone, in an experiment Kopec (1922) extirpated the brain of a caterpillar and demonstrated that this prevented pupation. He suggested that to a humoral factor produced by the brain was responsible for this phenomenon. This discovery heralded the study of hormonal regulation of metamorphosis in insects. Wigglesworth (1936) later described the secretion of a hormone that prevents metamorphosis from a pair of glands, the corpora allata (CA) attached to the base of the brain; he called it the "status quo" or juvenile hormone (JH). A few years later, Fukuda (1944) described the prothoracic glands as the source of the moulting hormone or ecdysone, which was later characterized as a steroid hormone (Butenandt and Karlson, 1954). At about the same time, the physiology of JH was elegantly worked out by Williams (1952). The chemical characterization of its structure, however, eluded researchers until 1967 when Roller and co-workers finally showed that it was a sesquiterpene, epoxy farnesoic acid methyl ester.

At this time, Williams (1967) made the now famous statement, "Third Generation Pesticides" in describing the use of JHs as environmentally safe control agents to which the insect will be unable to develop resistance. The development of highly potent synthetic analogues of JH, which were several fold more active than the native hormone, gave credence to William's claim (Henrick *et al.*, 1973).

Juvenile hormone has a unique terpenoid structure and is the methyl ester of epoxy farnesoic acid. This sesquiterpene exists in at least six different forms. JH III is the most common type and is present in most insects. Five different JHs, JH 0, I, II, III, and 4-methyl-JH I, have been described in lepidopterans. The bis-epoxide of JH, JH B_3, is found along with JH III in higher Diptera (Cusson and Palli, 2000). Various studies show that JH III is the main JH in most insects whereas JH I and II are the principal ones in Lepidoptera. Biosynthesis of JH proceeds through the mevalonate pathway with acetyl coenzyme A (acetyl CoA) serving as the building block. Propionyl CoA is used wherever ethyl side chains occur. Epoxidation and esterification are the terminal steps in the biosynthetic process (Schooley and Baker, 1985 and Brindle *et al.*, 1987).

Structures of Juvenile Hormone Analogues

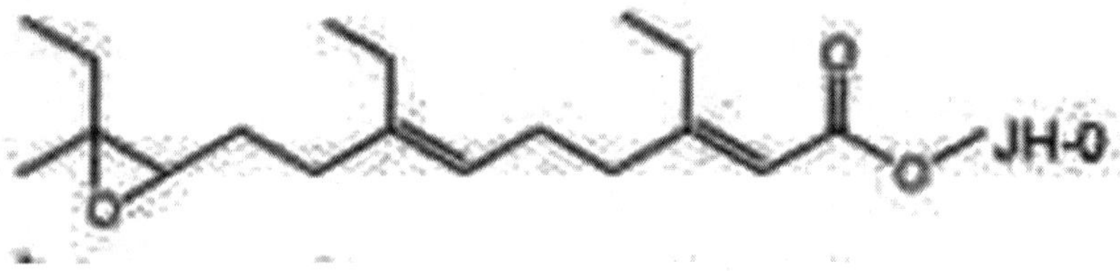

Chemical Structure of JH 0

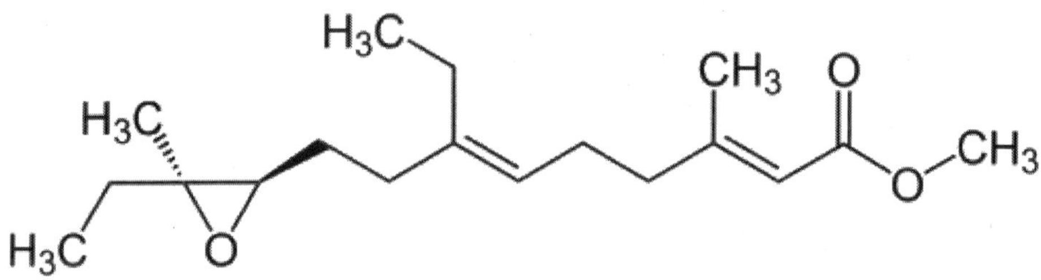

Chemical Structure of JH I.

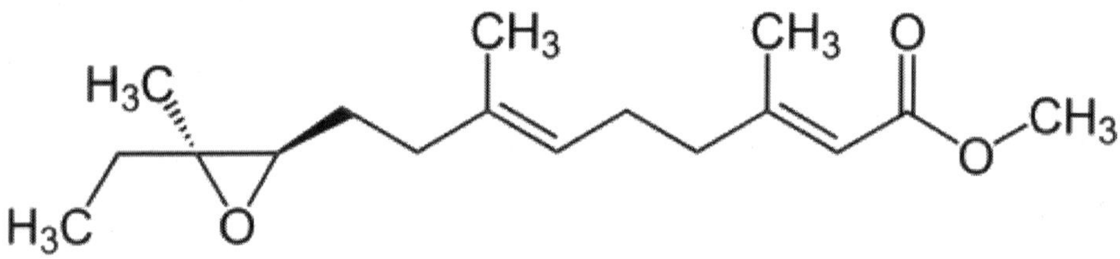

Chemical Structure of JH II.

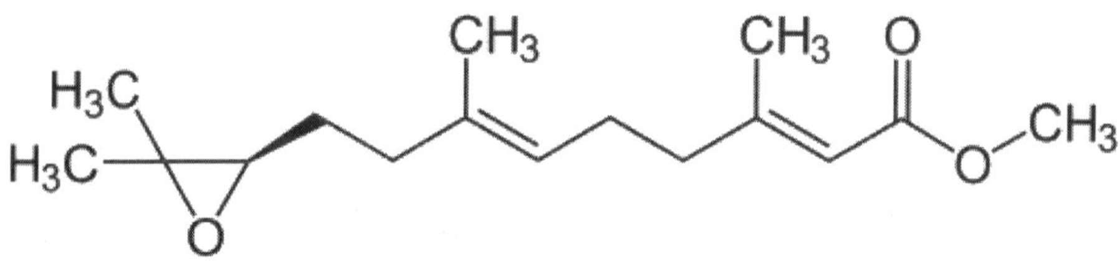

Chemical Structure of JH III.

JH secretion by the corpus allatum is regulated by two neurohormones, the allatotropins that stimulate secretion and the allatostatins that inhibit production. Severing the nerve connections or extirpating the neurosecretory cells results in the loss of control (Tobe and Stay, 1985; Tobe *et al.*, 1985 and Lee *et al.*, 2002). JH is cleared from the hemolymph by JH esterase, which selectively cleaves the methyl ester inactivating the hormone (Wroblewski *et al.*, 1990 and Feng *et al.*, 1999). JH being highly lipophilic is made soluble by JH binding proteins (JHB) to facilitate transportation to the target sites as well as protect it from degradation from nonspecific esterases (Goodman, 1990). JH is perhaps one of the most pleiotropic hormones known and functions in various aspects of metamorphosis, reproduction and behaviour (Riddiford, 1994, 1996, 2008; Wyatt and Davey, 1996; Cusson and Palli, 2000; Palli and Retnakaran, 2000; Hiruma, 2003 and Riddiford *et al.*, 2003). The major function of JH is the maintenance of the larval status or the so-called juvenilizing effect. During the last larval instar in Lepidoptera there is an absence of JH during the commitment peak of 20-hydroxyecdysone which results in the reprogramming of metamorphosis towards pupation. In the absence of JH, this ecdysone peak induces the expression of the broad complex gene (BrC or Broad), which is a transcription factor that initiates the larval-pupal transformation through several micro RNAs (miRNAs) (Zhou and Riddiford, 2002 and Sempere *et al.*, 2003). In adults, JH secretion is responsible for yolk protein (vitellogenin) synthesis and transport into the ovaries (Wyatt and Davey, 1996). In addition, JH is also responsible for adult diapause where the ovaries do not develop due to the absence of JH and this effect can be reversed by applying JH. JH has also been shown to play a role in caste determination, pheromone production, polyphenism, dispersal polymorphisms, flight, migration, antifreeze protein production, female sexual behavior, male accessory gland secretion *etc.* (Denlinger, 1985; Wyatt and Davey, 1996; Nijhout, 1999; Gilbert *et al.*, 2000; Zera and Cisper, 2001; Wheeler and Nijhout, 2003; Goodman and Granger, 2005; Raikhel *et al.*, 2005 and Zhan *et al.*, 2011).

The mode of action of JH at the molecular level has not been completely understood at present. Three types of action have been hypothesized and varying degrees of support for each of these modes have been reported. First is the direct action where it directly induces the secretion of a product (Iyengar and Kunkel, 1955 and Feng *et al.*, 1999). Second is the indirect action where it modulates the action of 20E (Dubrovski *et al.*, 2000 and Zhou and Riddiford, 2002). In these two instances, the action is at the genomic level where JH has to bind to nuclear receptors to evoke its action. The third involves the role of JH in the transport of vitellogenin into the oocytes through a membrane receptor (Sevala and Davey, 1989 and Wyatt and Davey, 1996). JH acts on the follicle cells of the ovary and makes them to contract, causing large spaces to appear between the cells, this process has been termed "patency." This intercellular space formation (patency) permits the vitellogenin from the hemolymph to enter the ovary and gain access to the surface of the oocyte and subsequently enter these cells. This phenomenon was originally described in *Rhodnius prolixus* but was later shown in many other species including locusts (Wyatt and Davey, 1996). This membrane-bound receptor appears to be coupled to a G-protein and the signal is transduced through a protein kinase system activating an ATPase that permits, among other things, the transport of vitellogenin into the cells and so the search goes on to unambiguously identifies the nuclear and membrane receptors of JH. Characterizing such a receptor would open the door for a more rational approach to developing target specific control agents.

The search to apply knowledge of JHs to the development of effective insecticides has since been limited to the area of JH analogues (Matolcsy *et al.*, 1988). Compounds that mimic the action of natural juvenile hormones are called as juvenile hormone analogues (JHAs) or said to be active mimics of JHs (Matolcsy *et al.*, 1988). JHAs can function as agonists or antagonists or a mixture of both with natural JHs (Kramer and Stall, 1981) and generally show the highest toxicity when applied at the onset of metamorphosis (Wilson, 2004). JHAs interfere with important biochemical mechanisms such as the secretion and transportation of natural JHs from the secretory site to the target site, degradation, excretion and feedback control (Retnakaran *et al.*, 1985). JHAs act at the genetic level and are associated with transcription of mRNA (Coudron *et al.*, 1981). Hence, their biological effects are very complex, and vary from one analogue to another.

Since, the early 1970s, numerous analogues of JH (juvenoids) have been tested for insecticidal activity (Sehnal, 1976). Most of the early analogues resemble JH in their basic terpenoid structure. The first juvenoids were farnesol and farnesal isolated from insects themselves (Dhadialla *et al.*, 1998). The 'paper factor' (Slama and Williams, 1965), now called juvabione, represents a group of hormone mimics present in a variety of plants where they may function as defensive mechanisms against herbivorous insects (Dhadialla *et al.*, 2005). Two very active JH analogues, methoprene and hydroprene lack the epoxide function present in JH. Some other highly active compounds with less apparent similarity to JH (aromatic non-terpenoidal JH analogues) like fenoxycarb, pyriproxyfen and diofenolan have been synthesized (Dhadialla *et al.*, 1998). JHAs differ widely from the commonly used insecticides, as they exert their insecticidal effects through their influence on development, metamorphosis and reproduction of the target insects by disrupting the normal activity of the endocrine system (Smet *et al.*, 1989, 1991 and Oberlander *et al.*, 1997).

One of the JHA, methoprene was found to be effective against the ontogeny of stored product pests like red flour beetle, *Tribolium castaneum* (Dyte, 1972; Loschiavo, 1975, 1976; Mian and Mulla, 1982a; Smet *et al.*, 1989; Arthur *et al.*, 2001; Parthasarathy and Palli, 2009; Wijayaratne *et al.*, 2010, 2012a, 2012b and Arthur and Fontenot, 2012); confused flour beetle, *T. confusum* (Loschiavo, 1975; El-Sayed, 1984 and Arthur *et al.*, 2011); *R. dominica* (Mian and Mulla, 1982a; Bengton, 1987; Cogburn, 1988; Daglish and Samson, 1990; Samson *et al.*, 1990; Kostyukovsky *et al.*, 2000; Arthur, 2003 and Athanassiou *et al.*, 2011a); *C. cephalonica* (Ambika and Abraham, 1982; Deb and Chakravorty, 1985; Tripathi and Tiwari, 2013, 2014), almond moth, *E. cautella* (Lazarovici *et al.*, 1984; Shaaya and Pisarev, 1986 and Chandra and Tiwari, 2013) and sawtoothed grain beetle, *Oryzaephilus surinamensis* (McGregor and Kramer, 1975; Loschiavo, 1975, 1976; Mian and Mulla, 1982a, 1982b and Athanassiou *et al.*, 2011a). Pyriproxyfen also affected the developmental stages of *O. surinamensis* (Arthur *et al.*, 2009); *T. castaneum* (Kostyukovsky *et al.*, 2000; Arthur, 2003 and Arthur *et al.*, 2009); *T. confusum* (Loni *et al.*, 2011 and Arthur, 2003); *R. dominica* (Kostyukovsky *et al.*, 2000); rice weevil, *Sitophilus oryzae* (Kostyukovsky *et al.*, 2000); yellow meal worm, *Tenebrio molitor* (Aribi *et al.*, 2006); sun pest, *Eurygaster integriceps* (Mojaver and Bandani, 2010); cigarettle beetle, *Lasioderma serricorne* (Arthur *et al.*, 2009); *Lipaphis orsini* (Liu and Chen, 2001); onion thrips, *Thrips tabaci* (Liu, 2003), *Bicyclus anyana* (Steigenga *et al.*, 2006); soybean aphid, *Aphis glycines* (Richardson and La-gos, 2007) and *P. interpunctella* (Arthur *et al.*, 2009 and Ghasemi *et al.*, 2010). Hydroprene also influenced the developmental stages of *C. cephalonica* (Samui *et al.*, 1981, 1982; Deb and Chakravorty, 1982 and Bhargava and Devaraj Urs, 1992, 1993); *E. cautella* (Nickle, 1979; Bell and Edwards, 1999), *T. casteneum* (William and Amos, 1974; Loschiavo, 1975; Arthur, 2003; Arthur *et al.*, 2009 and Parthsarthy and Palli, 2009); *T. confusum* (Loschiavo, 1975, 1976; El-Sayed, 1988; Smet *et al.*, 1989 and Arthur, 2003); merchant grain beetle, *O. mercator* (Loschiavo, 1975); *O. surinamensis* (Loschiavo, 1975 and Arthur *et al.*, 2009); *P. interpunctella* (McGregor and Kramer, 1975; Arbogast *et al.*, 2002; Mohandas *et al.*, 2006a, 2006b and Arthur *et al.*, 2009); *S. oryzae* (Gupta and Mkhize, 1983) and *E. kuehniella* (Hong, 1975 and Tan and Tan, 1980).

Role of JHAs on the growth duration of insects have also been reported in different insect species like *C. cephalonica* exposed to methoprene (Deb and Chakravotry, 1985); resistant strains of *T. castaneum* treated with methoprene (Kostyukovsky *et al.*, 2000); *P. interpuctella* following exposure of pyriproxyfen (Ghasemi *et al.*, 2010); *E. integriceps* treated with pyriproxyfen (Mojaver and Bandani, 2010); cowpea weevil, *Callosobruchus maculatus* exposed to hydroprene (Rup and Chopra, 1984). Adult longevity has also been found to be influenced by the exposure of IGRs as observed in case of *T. molitor* treated with diflubezuron (Soltani, 1984); *C. cephalonica, C. maculatus* and *T. castaneum* exposed to methoprene and hyroprene (Murali Baskaran and Janarthanan, 1991); *R. dominica, S. oryzae* and *T. castaneum* following treatment with methoprene and pyriproxyfen (Kostyukovsky *et al.*, 2000); *P. interpuctella* treated with pyriproxyfen (Ghasemi *et al.*, 2010); *E. integriceps* exposed to pyriproxyfen (Mojaver and Bandani, 2010); *T. castaneum and T. confusum* following treatment with methoprene (Tucker *et al.*, 2014b).

Compounds developed to disrupt metamorphosis ensure that no reproductive adults are formed. Those that specifically interfere with reproduction may include the development of adults with certain morphogenetic abnormalities that reduce their reproductive potential. Adults may be sterile or possess abnormally developed genitalia, which hinders the mating process or the capacity to produce fertile offspring.

Fecundity is the total number of eggs produced by a female (Southwood, 1978) whereas fertility is the total number of live eggs (Southwood, 1978) or the number of hatched larvae (Park *et al.*, 1958). Fecundity and fertility are affected by the number of spermatophores and by the frequency of mating (Butlin *et al.*, 1987 and Walker and Allen, 2010). In some butterflies, males transfer sodium to the female during copulation, and this increases fertility (Pivnick and McNeil, 1987).

Juvenile hormone affects spermatogenesis (Koeppe *et al.*, 1985 and Ashok, 2000), production of male-accessory gland proteins (Gillott and Gaines, 1992 and Gillott, 1996), communication for mating (Teal *et al.*, 2000) and mating (Segura *et al.*, 2009). Removal of corpora allata (reduction of JH) during the pre-adult stages increases spermatogenesis in *Periplaneta americana* (Blaine and Dixon, 1970). The effects of JHAs on spermatogenesis vary from species to species. For example, JHAs reduce spermatogenesis in *R. prolixus* (Hemiptera: Reduviidae) (Dumser and Davey, 1974), but increase spermatogenesis in *Pterostichus nigrita* (Coleoptera: Carabidae) (Ferenz, 1963). Developing male accessory glands in *Oncopeltus fasciatus* Dallas (Hemiptera: Lygaeidae) larvae are more sensitive to JH than are the germinal tissues. In this species, JH has to be present during the fourth instar larvae for successful maturation of accessory glands and cannot be replaced at a later stage, proposing "the existence of a JH-dependent juvenile phase" (Koeppe *et al.*, 1985). In *Spodoptera mauritia* (Boisduval) (Lepidoptera: Noctuidae), treatment with an ecdysone analogue increases spermatogenesis (Benny and Nair, 2006). In *T. castaneum*, JH deficient males have a reduction in mating behavior and sperm transfer. Furthermore, mating of such males with normal females reduces fecundity (Parthasarathy *et al.*, 2009). There is information about the role of JH in reproduction of female insects. JH stimulates vitellogenesis in the fat body in cockroach *Blattella germanica* (L.) (Dictyoptera: Blattellidae) (Comas *et al.*, 1999). In *Drosophila*, both JH and ecdysteroids are necessary for vitellogenesis (Bownes, 1989 and Richard *et al.*, 2001). In *Blaberus discoidalis* (Dictyoptera: Blaberidae), JH induces ovarian protein synthesis (Keeley and McKercher, 1985). In female *T. castaneum*, JH increases vitellogenesis (Parthasarathy *et al.*, 2010b), and ecdysteroid stimulates growth of ovaries and maturation of primary oocytes (Parthasarathy *et al.*, 2010a). In the stick insect *Dixippus morosus* (Orthoptera: Phasmidae), development of eggs and oviducts is decreased when the late-instar nymphs are provided with JH (Koeppe *et al.*, 1985). In female *P. americana*, JH applied at the nymphal stage reduces the development of the colleterial gland (Bodenstein and Sprague, 1959).

Effects of juvenile hormone analogues on the development of male and female reproductive systems, subsequently affects their reproduction. Exposing male or female larvae of Western spruce budworm, *Choristoneura occidentalis* Freeman (Lepidoptera: Tortricidae) to methoprene (ZR-515) or hydroprene reduces adult fecundity (Robertson and Kimball, 1979). In some cases, the effects of JHAs depend on the life stage. Methoprene applied to codling moth, *Cydia pomonella* fifth instar female larvae aged 2 and 6 days, reduces fecundity. However, there is no reduction in fecundity when 4-day-old larvae are exposed to methoprene (Brown and Brown, 1982). Sex-specific effects of methoprene are detected in certain species. When male and female larvae of the summer fruit tortrix moth, *Adoxophyes orana* (Lepidoptera: Tortricidae) was exposed to JHA (Ro 10-3108/018), fecundity of adults was reduced. In this case, females were affected more than males, as indicated by the number of eggs produced when paired with an untreated mate (Schmid *et al.*, 1978).

Pyriproxyfen has been reported to affect the reproductive behaviour of *P. interpucntella* (Ghasemi *et al.*, 2010); *E. integriceps* (Mojaver and Bandani, 2010 and Amiri *et al.*, 2012) and *T. confusum* (Loni *et al.*, 2011). Methoprene has also influenced reproductive changes in *T. castenium* and *T. confusum* (Williams and Amos, 1974; Loschiavo, 1976; Shanthy *et al.*, 1995 and Arthur, 2003); *E. cautella* (Nickle, 1979); *R. dominica* (Daglish and Pulvirenti, 1998; Chanbang *et al.*, 2008 and Athanassiou *et al.*, 2011b); book lice, *Liposcelis bostrychophila, L. entomophila, L. paeta, L. decolor* and *Lepinotus paeta* (Athanassiou *et al.*, 2010) and rice leaf roller, *Cnaphalocrocis medinalis* (Sun *et al.*, 2013).

IGRs have been found to influence the gonadial biochemistry in a number of insect species as observed in case of European mole cricket, *Gryllotalpa gryllotalpa* treated with FMC-23509 (CRD-9499), a juvenoid, (Mandal, 1982); green jewel bug, *Chrysocoris stolli* exposed to a juvenile hormone analogue (Saha *et al.*, 1985);

M. domestica exposed to different IGRs like dimilin, BAY SIR-8514 and altosid (Bakr, 1986); *Manduca sexta* exposed to 20-hydroxyedysone (Caglayan, 1990); the common house mosquito, *C. pipiens* (El-Bassal, 1993); pink bollworm, *Pectinophora gossypiella* and cotton spotted bollworm, *Earias insulana* (Anan *et al.*, 1993) and African cotton leafworm, *Spodoptera littoralis* (Abdel-Aal, 2002); *C. pomonella* (L.) treated with diflubezuron (Soltani and Soltani-Mazouni, 1997); lawn armyworm, *S. mauritia* treated with hydroprene (Zeenath and Nair, 1994); desert locust, *S. gregaria* treated with pyriproxyfen (El-Sherif, 1995) and lufenuron (Bakr *et al.*, 2007); *S. littoralis* exposed to pyriproxyfen (Shaurub *et al.*, 1998) and Tebufenozide (Abd El-Mageed, 2008); black cutworm, *Agrotis ipsilon* treated with pyroproxyfen (El-Sheikh, 2002); *Parasarcophaga aegyptiaca* exposed to pyriproxyfen and hexaflumuron (Assar, 2004); silkworm, *Bombyx mori* exposed to pyriproxyfen (Etebari *et al.*, 2007); *E. kuehniella* treated with tebufenozoide (an ecdysteroid agonist) (Khebbeb *et al.*, 2008); *P. interpunctella* following exposure with pyriproxyfen (Ghasemi *et al.*, 2010); *M. domestica* following treatment with pyriproxyfen (Assar *et al.*, 2010); oriental leafworm moth, *S. litura* treated with chlorfluazuron (Perveen and Miyata, 2000 and Perveen, 2012). The above investigations suggest that IGRs influence the internal biochemistry of the insects that may disrupt their metabolic flux which impairs their physiological fitness that contribute to the lethal action of IGRs.

Fenoxycarb, $C_{17}H_{19}NO_4$ (Ro- 135223, ACR -2907B, ACR-2913A, NRK-121, CGA-114597; commercial trademarks: Insegar™, Preclude™, Precision™, Award™, Logic™, Polyon™, Torus™, Pictyl™, Varikill™, Dicare™, Helgar™ and Comply™) (Dhadiall *et al.*, 2012) molecule has a mode of action identical to the structural analogues despite being structurally dissimilar to insect juvenile hormone (Dorn *et al.*, 1981 and Grenier and Grenier, 1993), was discovered in 1981 and introduced by Maag (The British Crop Protection Council, 1987). It is a non neurotoxic carbamate and the first JHA compound introduced to control agricultural pests (Miyamoto *et al.*, 1993). It has shown JHA activities against insects but with the advantage of being more stable and their activity more persistent. Fenoxycarb was the first JHA compound introduced to control agricultural pests (Miyamoto *et al.*, 1993.) and has shown JHA activities against a variety of insect orders including Lepidoptera, Coleoptera, Homoptera, Dictyoptera, Diptera and Orthoptera (Maner *et al.*, 1980). Because of its high activity and foliar stability, fenoxycarb is especially effective for the control of lepidopteran pests in orchard and vine crops, *e.g. C. pomonella* and light brown apple moth, *Epiphyas postvittana* (Grenier and Grenier, 1993). This pesticide is generally useful for control of fire ants, fleas, mosquitoes, cockroaches, moths, scale insects, and insects attacking vines, olives, cotton and fruit (Schaefer *et al.*, 1987; Reid, 1990; The British Crop Protection Council, 1991; Extension Toxicology Network, 1993 and Miyamoto *et al.*, 1993). It is also used to control these pests on stored products, and is often formulated as a grit or corncob bait.

Due to its effect on many species of several insect orders, specificity and low toxicity it is a desirable JHA for the use as postharvest protectant of stored agricultural commodities (Moreno *et al.*, 1992). Kramer *et al.* (1981) tested Ro-135223 against several coleopterans (*S. oryzae, S. granarius, S. zeamais, T. confusum, T. castaneum, O. surinamensis, R. dominica, Cryptolestes pusillus* and *Trogoderma variabile*), and lepidopterans (*P. interpunctella, Sitotroga cerealella* and *E. cautella*) pests of stored products with good results. Moreno *et al.* (1992) observed the effect of topical application of fenoxycarb on *E. kuhniella* and found the supernumerary larvae, larval-pupal intermediate and morphogenetic abnormalities leading to very low (10-0 per cent) adult emergence. It also exhibits some non JHA-specific effects on many insects (Retnakaran *et al.*, 1985). Its application also found to be effective against the ontogeny of *T. castaneum* (Edwards *et al.*, 1991; Thind and Edwards, 1986 and Cogburn, 1988); *O. surinamensis* (Kramer *et al.*, 1981; Thind and Edwards, 1986 and Edwards *et al.*, 1987, 1991); *S. oryzae* (Edwards and Short, 1984; Kramer *et al.*, 1981, 1985 and Ammar, 1990); *E. Kuehnilla* Zell. (Moreno *et al.*, 1992); *R. dominica* (Kramer *et al.*, 1981; Edwards *et al.*, 1987, 1991 and Samson *et al.*, 1990); *S. cerealella* (Kramer *et al.*, 1981; Cogburn, 1988; Eisa, 1992 and Letellier *et al.*, 1995); *Chrysoperla rufitebris* (Liu and Chen, 2001); *S. granarius* (Edward *et al.*, 1991); *Choristoneura fumiferana* (Mulye and Gordon, 1989; Hicks and Gordon, 1992); *Heliothius virescens* (Mauchamp *et al.*, 1989) and *Ostrinia nubilalis* (Gadenne *et al.*, 1990).

Grenier and Grenier (1993) in their extensive review of fenoxycarb summarized its effects on mites (Acari), sucking insects (Heteroptera and Homoptera) stored product pests, cockroaches (Blattaria), termites (Isoptera), ants (Hymenoptera: Formicidae), mosquitoes (Dipera: Culicidae), fleas (Siphonaptera), moth (Lepidoptera), especially leafrollers (Tortricidae), and some acridids (Orthoptera: Acrididae).

Plant-sucking insects are susceptible to fenoxycarb. The compound practically controlled the woolly aphid, *Eriosoma lanigerum* (Homoptera: Aphidae) (Nicholas *et al.*, 2005). Fenoxycarb is efficacious against scale insects such as *Phenacocuus pergandei* (Homoptera: Pseudococcidae) (Zhou *et al.*, 2008) and *Diaspidiotus perniciosus* (Diaspididae) (Sazo *et al.*, 2008). Conventional insecticides were found to be more efficacious against *Cacopsylla melanoneura* (Homoptera: Psyllidae) than fenoxycarb (Baldessari *et al.*, 2010). The latter was less efficacious than pyriproxyfen against the thrips, *Frankliniella occidentalis* (Thysanoptera: Thripidae) (Ludwig and Oetting, 2001). The red imported fire ant, *Solenopsis invicta* (Hymenoptera: Formicidae) is susceptible to bait (Logic™) containing fenoxycarb (Williams *et al.*, 2001) like to the bait containing methoprene.

Hoehn and Wildbolz (1992) reported that in Swiss orchards, spraying fenoxycarb (formulated as wettable powder containing 25 per cent a.i.) effectively controlled the codling moth, *C. pomonella* (Lepidoptera: Tortricidae), a pest in apple and pear orchards. Spraying fenoxycarb against the codling moth yielded beneficial side effects for control of other lepidopteran orchard pests like the small ermine moth, *Yponomeuta malinellus* (Yponomeutidae), the casebearer moth, *Coleophora* spp. (Coleophoridae), and the non-lepidopteran apple leaf sucker, *Cacocpsylla mali* (Homoptera: Psyllidae). Good ovicidal control of the small fruit tortrix, *Grapholita labarzewskii* (Lepidoptera: Tortricidae) was also observed. Pasquier and Charmillot (2004) reconfirmed that fenoxycarb is highly efficacious against a susceptible population of diapausing larvae of the codling moth. However, this pest has developed high level of resistance to fenoxycarb. Good ovicidal action of fenoxycarb (Insegar™ DG formulation) on the small fruit tortrix moth was reconfirmed by

Charmillot *et al.* (2007). Malaysian farmers effectively used fenoxycarb (Dicare™) against the diamond black moth, *Plutella xylostella* (Lepidoptera: Yponomeutidae: Plutellinae), a pest of cabbage (Mazlan and Mumford, 2005). Zero day old pupa of the beet armyworm, *Spodoptera exigua* (Lepidoptera: Noctuidae) was highly susceptible to fenoxycarb that inhibited adult emergence but susceptibility decreased with advancing age of the pupae (Kim *et al.*, 2000).

Fenoxycarb (Torus™) exerted strong residual activity against cat flea, *Ctenocephalides felis* (Siphonaptera: Pulicidae) in topsoil (Rajapakse *et al.*, 2002). This residual activity was similar to that of pyriproxyfen but superior to that of slow release methoprene.

Recently, Kavallieratos *et al.* (2012) found that fenoxycarb as well as a combined insect growth disrupter *i.e.* fenoxycarb + lufenuron (a CSI) sprayed on grains, caused slight or moderate direct mortality of adults of the stored product pests, *Prostephanus truncantus* and *R. dominica* (both Coleoptera: Bostrichidae) and their progeny production was suppressed by over 93 per cent in *P. truncates* and 88.6 per cent in *R. dominica.*

Elvira *et al.* (2010) reported an intresting aspect of utilization of fenoxycarb which produced multiple nucleopolyhedrovirus in beet armyworm, *S. exigua*, that markedly enhanced in supernumerary sixth instar larvae. The authors suggested the use of this approach for mass production of viral occlusion bodies.

Fenoxycarb and pyriproxyfen, a fenoxycarb derivative in which part of the aliphatic chain has been replaced by pyridyl oxyethylene, are the JHAs with high insecticidal activities (Riddiford, 1994). Fenoxycarb mimics the action of the juvenile hormones on a number of physiological processes, such as moulting and reproduction. It can kill eggs and larvae of numerous insect species (s *et al.*, 1987). Since, the egg is not usually exposed to high levels of JH until about halfway through embryonic development, its development is halted and the egg will not hatch. High levels of JHAs, when applied to later instars, cause the adult insect to maintain larval characteristics and these insects generally cannot reproduce. Another property of these compounds is that, in adults, they disrupt normal reproductive physiology and act as a method of birth control.

In the European corn borer, *Ostrinia nubilalis*, application of fenoxycarb during the second to fourth larval instars had no effect on the duration of these instars. However, the duration of the resulting fifth instars increased significantly (Gadenne *et al.*, 1990). Applications of fenoxycarb in the fifth instar produced different effects, depending upon the dose and the timing of application, which resulted in production of supernumerary or permanent larvae, or of larval-pupal intermediates (Gadenne *et al.*, 1990).

The vapour pressure of fenoxycarb is low (4.6 x 10^{-5} Pa m^3/mol), which indicates that it does not have a strong tendency to volatilize into the atmosphere. Aside from drift that may occur with spray applications, fenoxycarb is not expected to be found in the air. Fenoxycarb is relatively unstable with a half-life of one day in soil and five hours in direct sunlight. Fenoxycarb has relatively low water solubility and readily adsorbs to soil surfaces. The compound has a low potential for leaching from the soil and has a moderate to strong tendency for soil binding (U.S. Environmental Protection Agency, 1986). These characteristics of fenoxycarb in soil indicate that it is unlikely to contaminate groundwater. It is relatively quickly degraded in soil by microbial action, and has an overall reported field dissipation half-life of 14-45 days. However, fenoxycarb is stable to photodegradation in most viable soil types (ARS Pesticide Properties Database, 1995). The photolysis half-life for fenoxycarb in soil under artificial light was determined to follow biphasic kinetics with a primary half-life ranging from 6 to 14 days and a secondary half-life ranging from 73 to 1351 days (Thede, 1995). Thede (1995) conducted a field study for twelve months at two doses of ^{14}C-A ring fenoxycarb to determine metabolic degradation rates in a Maryland limekiln sandy loam soil. Aerobic kinetic incubations were dosed at 0.122 ppm and the bulk incubations were dosed at 9.16 ppm. Degradation of fenoxycarb for the aerobic and anaerobic kinetic phases followed biphasic kinetics. The primary half-life for the aerobic biphasic kinetic incubation was determined to be 6.7 h and the secondary half-life value was calculated to be 8.2 months. The aerobic/anaerobic half-life was calculated to be 113.6 days. By the twelfth month, volatiles (CO_2) accounted for around 42.5 per cent of total dose in the aerobic incubations and 32.1 per cent of total dose in the aerobic/anaerobic kinetic incubations. Multiple degradation pathways appear to contribute to the conversion of fenoxycarb to CO_2 and there was evidence that some radiolabeled carbon from the fenoxycarb was incorporated into the soil biomass. Evidence indicates that fenoxycarb rapidly degrades to several minor components that do not accumulate over time. The degradates in turn rapidly degrade into CO_2 (Thede, 1995). Terrestrial field studies of ^{14}C-fenoxycarb on bare ground soil in California showed that most residues remain in the first 3 inches of soil. Degradation of parent followed biphasic kinetics with primary and secondary half-lives ranging from 3.07-5.11 days and 13.7-44.90 days, respectively (McDonald, 1995). Schuster and Goff (1995) determined that fenoxycarb was immobile and dissipate rapidly and no residues were found in any other soil layer beyond the zero to 6-inch soil horizon. It is stable to hydrolysis in water at pH 3-9 and temperatures up to 50°C. Thede (1995) reported that the half-life of fenoxycarb in aqueous buffer was 1,406, 3,136 and 4,101 days at pH 5, 7 and 9 respectively. Based on these data, hydrolysis would not be expected to be a major degradation pathway for fenoxycarb. Fenoxycarb is thus much less stable to photolysis than it is to hydrolysis in water at environmental pH range. The aqueous photodegradation of fenoxycarb was determined to be 18-23 days at 25°C under artificial sunlight at approximately 1 ppm and at a hydrolytically stable aqueous buffered pH of 7 (Thede, 1995). Fenoxycarb is stable in water over a range of temperature (10-38°C) and pH (6.5-10) and is expected to break down relatively quickly in plants (Kidd and James, 1991). It has very low toxicity to mammals via ingestion. The oral LD_{50} of the compound is greater than 10,000 mg/kg in rats (Keller, 1982). It is slightly toxic via the dermal route, with a reported dermal LD_{50} in the rat of greater than 2000 mg/kg. It is not a skin sensitizer in guinea pigs and causes only minimal eye irritation when applied to rabbits. The inhalation toxicity of fenoxycarb is moderate, with an acute inhalation LC_{50} in rats of greater than 0.480 mg/L (Kidd and James, 1991). It has low toxicity to birds and it's LD_{50} values are greater than 3000 mg/kg and 7000 mg/kg in mallard ducks and bobwhite quail, respectively

(Kidd and James, 1991). Fenoxycarb is considered moderately to highly toxic to fish with LC_{50} values ranging from 1.6 mg/L in rainbow trout to 10.3 mg/L in carp (Kidd and James, 1991). It is also considered highly toxic to the aquatic invertebrate *Daphnia* and affects growth and reproduction after chronic exposures to >1.6 ng/L (U.S. Environmental Protection Agency, 1986). However, when it was applied at rates ranging from 0.015 to 0.03 lbs/acre to ponds, the compound had no effect on a number of different invertebrates including cladocerans, copepods, ostracods and mayfly nymphs (Miura *et al.*, 1986). In one study, bluegill sunfish accumulated 20 times the amount of the compound's concentration in the water. Tissue residues of the pesticide quickly declined after the fish were placed in pesticide-free water (Schaefer *et al.*, 1987). In laboratory tests, contradictory results are obtained with the mosquito fish, *Gambusia affinis*. Miura *et al.* (1986) found no mortality with fenoxycarb doses from 0.0001 to 1.0 ppm while Tietze *et al.* (1991) reported LC_{50} value of 1.05 ppm after 24 hour exposure. Miura and Takahashi (1987) also reported that fenoxycarb treatment rates of 0.03 kg a.i./ha had no deleterious effects on planktonic organisms and aquatic beetles regularly found in mosquito breeding habitats. Fenoxycarb is practically non-toxic to honeybees (Kidd and James, 1991). The oral LC_{50} (24 hour) for adult honeybees is greater than 1000 ppm. Young bumble bee, *Bombus terrestris* colonies treated with 1:1 sucrose solution containing 0.20, 2.0 and 20 ppm fenoxycarb and pyriproxyfen exhibited no larval mortality and the brood developed normally (Wael *et al.*, 1995). Bee hazard is decreased when the pesticide is formulated as grit or corncob bait. However, it is recommended (Grenier and Grenier, 1993) that more care be taken by avoiding application during blossoming periods, cutting the grass on the ground under trees to remove all the flowers before treatment, and making applications outside normal flight hours of bees.

Although, sufficient knowledge exist on the effect of insecticidal agents (organochlorines, organophosphates, synthetic pyrethroids and natural plant products influencing ontogeny, as well as larval haemolymph and fat body biochemistry as well as gonadial biochemistry of *C. cephalonica* (Tiwari, 1987; Tiwari and Bhatt, 1987, 1988, 1989, 1992, 1993, 1994a, 1994b, 1994c, 1994d, 1994e, 1996, 1999, 2000a, 2000b; Tiwari and Tripathi, 2001, 2003, 2005, 2006; Pathak and Tiwari, 2010a, 2010b; Shukla and Tiwari, 2011a, 2011b, 2011c; Pathak and Tiwari, 2012 and Shukla and Tiwari, 2013) but scientific contribution in relation to juvenile hormone analogues (JHAs) influencing physiological *i.e.* ontogeny, growth duration, adult longevity, fecundity, fertility and sterility as well as gonadial and egg biochemistry of rice moth, *C. cephalonica* is completely wanting. The acquisition of such knowledge in this area becomes essential for a comprehensive appreciation of the physiological and ecological relationship that exists between this pest and its host material (stored cereals and cereal commodities). This knowledge in turn, is likely to generate new insights into divising ways and means for controlling *C. cephalonica,* by disrupting its metabolic frame work so that evolution of a new generation of this pest for the eventual establishment on stored cereals and cereal products can be considerably restricted. Hence, as an objective of such programme the present work for the first time, has been designed and conducted to examine into the impact of a juvenile hormone analogue (JHA) *i.e.* fenoxycarb on the ontogeny, growth duration, adult longevity, fecundity, fertility and sterility as well as on the gonadial and egg biochemistry of rice moth, *C. cephalonica*.

No doubt, the present study of the rice moth, *C. cephalonica* treated with fenoxycarb may provide valuable message to an eager entomologist who opts more about such JHAs induced changes on the ontogeny, growth duration, longevity of adults, fecundity, fertility, and sterility as well as on the gonadial and egg biochemistry of this lepidopterous insect, the latter results into perturbation of available metabolic reserves and thus impairs the physiological as well as reproductive fitness of the larva and thereby contributes to the lethal action of the fenoxycarb.

Chpater 2

Materials and Methods

Collection and Culture of Rice Moth, *Corcyra cephalonica*

Corcyra cephalonica Stainton adults were obtained from already existing laboratory stock culture maintained on normal dietary medium composed of coarsely ground jowar (*Sorghum vulgare*) mixed with 5 per cent (w/w) powdered yeast inside large glass containers (150 mm diameter, 200 mm height) at temperature 26 ± 1°C, relative humidity (R.H.) 93 ± 5 per cent and a light regime of 12 h light and 12 h darkness. Such a standard culture was maintained throughout the year.

Egg Laying and their Hatching

From the above culture whenever needed, newly emerged males and females were transferred to oviposition glass chambers (35 mm diameter, 200 mm height). Since *C. cephalonica* individuals do not feed during their adult stage, no food was provided to them during their confinement in these vessels. Eggs laid by the females were collected and then placed in glass chambers (consisting of 250 ml beakers) with the help of zero number camel hair brush for hatching.

Juvenile Hormone Analogue Used

Fenoxycarb ethyl[2-(4-phenoxy-phenoxy)-ethyl]carbamate, molecular formula-$C_{17}H_{19}NO_4$, a non terpenoid juvenile hormone analogue, P-686N, Lot-20071 used throughout the experiment, was obtained from AccuStandard, New Haven, CT 06513, USA.

Chemical Structure of Fenoxycarb

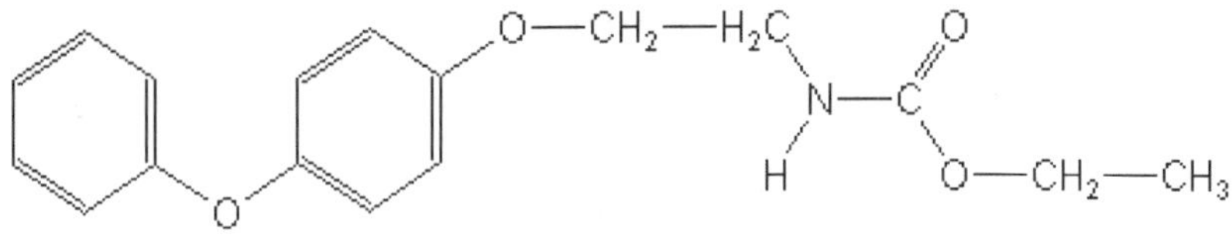

Fenoxycarb
Ethyl [2-(4-phenoxy-phenoxy)-ethyl] carbamate

Preparation of Different Concentrations of Fenoxycarb in Dietary Media

For this purpose, a stock solution of known concentration of JHA was prepared by dissolving it in acetone and then adjusted via serial dilutions to achieve its required concentrations. Now, required volume of different concentrations of fenoxycarb was thoroughly mixed with the required quantity of normal food (roughly ground jowar mixed with 5 per cent w/w yeast powder) to get different desired concentrations *i.e.* 0.001, 0.005, 0.01, 0.05, 0.10, 0.50 and 1.00 ppm of fenoxycarb in dietary media. This treated food was then air dried at room temperature to eliminate completely the acetone. For control purposes, the normal food was thoroughly mixed with a required volume of acetone similar to that of treated food and then air dried in the same way.

Evaluation of Toxicity of Fenoxycarb against the Developmental Stages of *C. cephalonica*

To evaluate the toxicity of various concentrations of fenoxycarb at different life stages of rice moth, four sets of experiments were designed. In each set, eight rearing chambers (consisting of 250 ml beakers) containing 50 g of food were used, out of which 7 rearing chambers contained 0.001, 0.005, 0.01, 0.05, 0.10, 0.50 and 1.00 ppm of JHA treated food separately, whereas one chamber contained normal food, serving as control. In the first set of experiment, freshly hatched larvae of *C. cephalonica* were allowed to feed on a normal dietary medium for exactly four days and on the fifth day, a group of 25 first instar larvae were transferred to each rearing chamber containing 50 g of dietary medium mixed and treated separately with different known concentrations of fenoxycarb and similarly in the 2nd, 3rd and 4th sets of experiment, the second instars, third instars and fourth instars larvae were kept, respectively. All the four sets of experiments were kept at the temperature, relative humidity and photophase, as mentioned earlier. On the completion of life-cycle, per cent adult emergence and pupal mortality were observed and on that basis per cent pupation and per cent larval mortality were calculated. The developmental course and external morphology of larvae, pupae and adults were also observed. Adult mortality was also noted up to 24 h of emergence. The corrected total mortality was calculated by Abbott's formula (1925), as:

$$\text{Corrected total mortality} = 100 \times \frac{\text{Per cent experimental mortality} - \text{Per cent control mortality}}{100 - \text{per cent control mortality}}$$

After completion of life-cycle, the remaining food was weighed to calculate the amount of fenoxycarb consumed per larva (μg/larva) at each concentration of fenoxycarb.

For this purpose they were kept in mating/oviposition chamber for 24 h and thenafter, the mated males and females were kept separately in the glass containers (35 mm diameter, 200 mm height). The growth duration (time taken from egg laying to adult emergence) and adult longevity (life-span) of males and females of *C. cephalonica* of each experiment were observed in both the cases *i.e.* in experimental as well as control sets.

Experiments were replicated six Experiments were replicated six times and values have been expressed as mean ± SEM. Student's t-test was applied to determine the significant differences between the treated groups and their control. LD_{50} values (μg/larva), 95 per cent confidence limits (lower and upper confidence limits) of LD_{50}, slope values and heterogeneity of fenoxycarb were calculated by Polo Plus, Probit and Logit Analysis, Version: 2.0, LeOra Software based on probit analysis (Finney, 1959).

Evaluation of Growth during and Longevity of Adults, Emerged from Treated Food

The adults, obtained from all the experimental sets of the previous experiments, were sexed just after their emergencetimes and the values have been expressed as mean ± SEM in days. Straight line regression equation was applied between different concentrations of fenoxycarb and their corresponding growth duration of males and females and longevity of adults to observe the significant correlation. (Sokal and Rohlf, 1969)

Evaluation of Egg Laying Capacity and Egg Hatchability

The adults emerged from experimental sets were immediately sexed and used for mating experiments. For this purpose, four types of crosses were made, as following:

1. Normal male x Normal female (Serving as control)
2. Treated male x Normal female
3. Normal male x Treated female
4. Treated male x Treated female

(Treated male and treated female refers to the male and female moths obtained from treated food).

These pairs were allowed to mate and lay eggs in the oviposition chambers (35 mm diameter, 200 mm height glass tubes) separately. The eggs laid were collected daily till the females died and the collected eggs were transferred to the hatching sites (glass petridishes of 100 mm diameter, 10 mm height) and allowed to hatch. The total number of eggs laid per pair (oviposition rate) was recorded and after their hatching the total number of hatched eggs (hatchability) was also recorded. Per cent hatching and per cent sterility were calculated on that basis. Corrected sterility was calculated by Chamberlain's formula (1962), as:

$$\text{Corrected sterility} = 100 \times \frac{\text{Per cent hatch in control} - \text{Per cent hatch in treated}}{\text{Per cent hatch in control}}$$

At each cross of each concentration in every instar, six pairs of males and females (each pair in separate mating/oviposition chamber) were kept for experimentation upto the concentrations where adults were available.

Experiments were replicated six times and the values have been expressed as the mean ± SEM. Student's t-test was applied to determine the significant differences between the treated groups and their control.

Dose Selection of Fenoxycarb for Biochemical Assay

For biochemical estimations, out of various concentrations of the fenoxycarb mentioned above, only such concentrations (0.001, 0.005 and 0.01 ppm) were selected, which allowed the larvae to survive and develop but caused considerable effect in the internal biochemistry of the gonads (testes and ovaries) and eggs (laid within 24 h of mating) that could be easily detected and assessed to prove the effectiveness of hormonal control measures against this lepidopterous pest.

Larval Rearing and their Treatment for Biochemical Estimation

As stated earlier, freshly hatched larvae were allowed to feed on a normal dietary medium (kept inside 250 ml beakers) for 4 days. On the 5th day, 25 first instar larvae were transferred to each similar rearing chambers containing dietary medium mixed with 0.001, 0.005 and 0.01 ppm concentrations of fenoxycarb and were allowed to complete their life-cycle and just after emergence, the males and females were collected and kept separately to avoid mating. Thereafter adults were sacrificed and their testes and ovaries were isolated for biochemical estimation. 25 larvae were also kept as control with each set of experiment.

Separation and Collection of Tissues/Eggs

For obtaining testes and ovaries, the newly emerged moths were anaesthetized with ether and were dissected under stereoscopic binocular microscope in 0.9 per cent sodium chloride solution. The gonads thus obtained, were washed with 0.9 per cent ice cold sodium chloride solution and then their biochemical estimations were made. Further, eggs laid within 24 h of mating by the females of all the four crosses (*i.e.* Normal male x Normal female, Treated male x Normal female, Normal male x Treated female and Treated male x Treated female) were collected separately for their biochemical estimations.

Biochemical Estimations

The adults, obtained from the first instar treated larvae, were used for biochemical studies. The entire programme of biochemical estimation includes the quantitative measurement of total protein, total free amino acids, nucleic acids and glycogen levels and the activity of acid and alkaline phosphatases in testes and ovaries of adults emerged from the first instar larvae exposed with sublethal concentrations (0.001, 0.005 and 0.010 ppm) of fenoxycarb. The biochemical estimation of eggs laid within 24 h of mating by the females of all the four crosses (*i.e.* Normal male x Normal female, Treated male x Normal female, Normal male x Treated female and Treated male x Treated female) were also carried out colorimetrically.

Total Protein

The total protein in gonads and eggs was measured according to the method of Lowry *et al.* (1951) using bovine serum albumin as standard.

Freshly separated tissues were homogenized in 10 per cent TCA (1 mg/ml, w/v) for gonads and 100 eggs/ml for eggs, in an electrical tissue homogenizer for 5 min and the homogenates were centrifuged at 6000 rpm for 20 min. The residue was washed with 5 per cent TCA and again centrifuged at the same speed for another 20 min. The residue thus obtained, was dissolved in required volume of 1 N NaOH solution. In order to develop colour, to 1.0 ml of the above mixture, 5.0 ml of freshly prepared alkaline copper solution (protein reagent) was added and the reaction mixture was left for 10 min at room temperature. Protein reagent was prepared by mixing 50.0 ml of reagent A (2 per cent sodium carbonate in 0.1 N NaOH) and 1.0 ml of reagent B (2 per cent sodium potassium tartarate, 1 per cent copper sulphate: mixed in the ratio of 1:1 at the time of experiment). Thenafter, 0.5 ml of Folin-ciocalteu's reagent (diluted in the ratio of 1:2 with distilled water at the time of the experiment) was added to the reaction mixture and mixed thoroughly. The blue colour was allowed to develop for 10 min. The absorbancy was measured at 600 nm against blank prepared simultaneously by using 1.0 ml of 1 N NaOH instead of 1.0 ml of tissue extract. The absorbancy was compared with that of varying concentrations of bovine serum albumin (20 mg/ml to 200 mg/ml). Appropriate calculations were made to compute the total protein content of the tissues. Results were expressed as µg protein/mg wet weight tissues for gonads and µg/egg for eggs.

Total Free Amino Acids

Estimation of total free amino acids was carried out according to the method of Spies (1957) using glycine solution as standard.

Freshly separated tissues were homogenized in 96 per cent ethanol in ratio of 10 mg/ml, (w/v) for gonads and 100 eggs/ml for eggs, in an electrical tissue homogenizer for 5 min and the homogenates were centrifuged at 8000 rpm for 20 min. The supernatants were subjected to total free amino acids estimation.

In order to develop colour, to 0.1 ml of supernatant, 0.1 ml of distilled water and 2.0 ml of ninhydrin reagent was added and mixed thoroughly. Ninhydrin reagent was prepared by mixing 1.0 g ninhydrin in 25.0 ml of absolute ethyl alcohol and 0.04 g of stannous chloride in 25.0 ml of citrate buffer pH 5.0. The reaction mixture was kept in boiling water bath for exactly 15 min. After cooling, 2.0 ml of 50 per cent ethyl alcohol was added to each tube. A violet colour developed. The absorbancy was measured with the help of spectrophotometer, systronics Digital Type 106 (MK II) at 575 nm against blank prepared in the similar way by using 0.1 ml of 96 per cent ethanol instead of 0.1 ml of supernatant.

The absorbancy was compared with that of varying concentrations of glycine solution (10 µg/ml to 100 µg/ml) in distilled water. Appropriate calculations were made to compute the total free amino acid levels. Results were expressed as µg/mg wet weight of tissues for gonads and µg/egg for eggs.

Estimation of Nucleic Acids

Deoxyribonucleic acid (DNA) and ribonucleic acid (RNA) levels in gonads and eggs of *C. cephalonica* were estimated according to the method of Schneider (1957). Diphenylamine reagent was used for DNA estimation while orcinol reagent was used for RNA estimation. Homogenates were prepared in an electrical tissue homogenizer in 5 per cent TCA, (1 mg/ml, w/v) for gonads and 100 eggs/ml for eggs, at 90°C, centrifuged at 5000 rpm for 20 min and supernatants were used for their estimations. Appropriate calculations were made to compute the DNA, RNA levels and the values were expressed as µg/mg wet weight of tissues for gonads and µg/egg for eggs.

Deoxyribonucleic Acid (DNA)

To 1.0 ml of supernatant, 1.0 ml of distilled water and 4.0 ml of freshly prepared diphenylamine reagent was added. Diphenylamine reagent was prepared by adding 1.0 g purified diphenylamine to 100 ml glacial acetic acid followed by addition of 2.5 ml conc. H_2SO_4. The reaction mixture was kept in boiling water bath for 10 min. A blue colour developed.

The absorbancy was measured at 600 nm against blank prepared in a similar way by using 1.0 ml of 5 per cent TCA instead of 1.0 ml of supernatant. The absorbancy was compared with that of different concentrations of calf thymus DNA.

Ribonucleic Acid (RNA)

To 1.0 ml of supernatant, 2.0 ml of distilled water and 3.0 ml of orcinol reagent was added. Orcinol reagent was prepared by adding 1.0 g orcinol in 100 ml concentrated hydrochloric acid (HCl) and 0.5 g ferric chloride. The reaction mixture was kept in boiling water bath for 20 min. A greenish blue colour developed.

The absorbancy was measured at 660 nm against blank prepared in a similar way by using 1.0 ml of 5 per cent TCA instead of 1.0 ml of supernatant. The absorbancy was compared with that of varying concentrations of yeast RNA.

Estimation of Glycogen

Glycogen was estimated according to the method of Van der Vies (1954) by using anthrone reagent. Glycogen, present in some microscopic form in tissues, yields an amorphous product on reacting with KOH. This produces a green-brown colour with anthrone reagent. Homogenates were prepared in 5 per cent TCA (10 mg/ml, w/v) for gonads and 100 eggs/ml for eggs and centrifuged at 6000 rpm for 20 min and supernatants were used for glycogen estimation. To 1.0 ml of this supernatant, 1.0 ml of 10 N KOH was added and the mixture was boiled for exactly 60 min. Excess alkali was neutralized with 0.5 ml glacial acetic acid and distilled water was added to make a final volume of 10 ml. To 1.0 ml of this mixture, 2.0 ml of freshly prepared anthrone reagent (2 mg anthrone/ml of 36 N H_2SO_4) was added. It was thoroughly shaken and heated in a boiling water bath for 10 min. A green-brown colour developed. The absorbancy was measured at 650 nm against blank, prepared simultaneously by using 1.0 ml of 5 per cent TCA instead of 1.0 ml of tissue filtrate.

The absorbancy was compared with that of varying concentrations (0.01 mg/ml to 0.4 mg/ml) of glycogen standard. Appropriate calculations were made to compute the glycogen levels and the values were expressed as µg/mg wet weight for gonads and µg/egg for eggs.

Estimation of Phosphatase Activity

Acid and alkaline phosphatase activities in gonads and eggs were determined according to the method of Andersch and Szcypinski (1947) as modified by Bergmeyer (1967) using p-nitrophenylphosphate as substrate. Both the phosphatases (acid and alkaline) catalyze the hydrolysis of p-nitrophenylphosphate into phosphoric acid and p-nitrophenol. Activity of phosphatases is directly proportional to the amount of p-nitrophenol formed. Tissue homogenates (2 mg/ml, w/v) for gonads and 100 eggs/ml for eggs were prepared in ice-cold 0.9 per cent sodium chloride solution and centrifuged at 5000 rpm for 20 min at -4°C in freezing centrifuge. The supernatants were used as enzyme source and were kept at -4°C before use. The absorbancy was compared with standard calibration curves which were drawn with the absorbancy measured from different concentrations of p-nitrophenol. Appropriate calculations were made to compute phosphatase activity. The activities of acid and alkaline phosphatases were expressed as µ moles substrate hydrolyzed/30 minutes/mg protein.

Acid Phosphatase Activity

Acid phosphatase activity was determined by adding 0.2 ml of enzyme source to 1.0 ml (previously incubated for 15 min at 37°C) of acid buffer substrate solution. Acid buffer substrate solution (pH 4.8) was prepared by dissolving 0.41 g citric acid, 1.125 g sodium citrate and 165 mg p-nitrophenylphosphate sodium salt in 100 ml of double distilled water. The reaction mixture was mixed thoroughly and incubated for 30 min at 37°C. Thenafter, 4.0 ml of 0.1 N NaOH (reaction inhibitor) was added to the incubation mixture. At the end of the reaction a yellow colour developed. The absorbancy was measured at 420 nm against blank prepared by using 0.2 ml of 0.1 N NaOH instead of enzyme sources.

Alkaline Phosphatase Activity

Alkaline phosphatase activity was determined by adding 0.10 ml of enzyme source to 1.0 ml (previously incubated for 15 min at 37°C) of alkaline buffer substrate solution. Alkaline buffer substrate solution (pH 9.5) was prepared by dissolving 375 mg glycine, 10 mg $MgCl_2.6H_2O$ and 165 mg p-nitrophenylphosphate sodium salt in 42.0 ml of 0.1 N NaOH and diluting the mixture with double distilled water to 100 ml. The incubation mixture was mixed thoroughly and incubated for 30 min at 37°C. 10.0 ml of the 0.02 N NaOH (reaction inhibitor) was added to the incubation mixture. A yellow colour developed at the end of the reaction. The absorbancy was measured at 420 nm against blank prepared by using 0.1 ml of 0.02 N NaOH instead of enzyme source.

Statistical Analysis

Results were expressed as mean ± SEM of six replicates. Significant differences between treatment groups in order to show dose-dependence, were determined by one way analysis of variance ($p < 0.001$ to $p < 0.05$) by using StatPlus® version 2009. Student's t test was applied to determine significant differences between the corresponding treated groups and the controls ($p < 0.001$ to $p < 0.05$).

Chapter 3

Observations

SECTION I A

Effect of Fenoxycarb on the Ontogeny of Rice Moth, *Corcyra cephalonica*

This section reveals the toxicological property of various concentrations of fenoxycarb on the developmental stages of the rice moth, *C. cephalonica* when exposed at its different instars *i.e.* I, II, III and IV instars (Tables 1–4, Figures 1–8).

Toxicity of Fenoxycarb on the Ontogeny of First Instar Larval Treatment (Table 1, Figures 1 and 2)

A significant larval mortality was obtained with the increase of fenoxycarb concentration in the diet. In case of control larvae, larval mortality was recorded to be 4.00 ± 1.03 per cent but at 0.001 ppm concentration of fenoxycarb larval mortality was observed to be 11.33 ± 1.23 per cent which increased to 100 per cent at 1.00 ppm concentration of fenoxycarb. As the fenoxycarb concentration increases a significant reduction in pupation and a significant enhancement in pupal mortality did occur. In case of control larval group, 96.00 ± 1.03 per cent pupation was observed that decreased to 88.67 ± 1.23 per cent at 0.001 ppm concentration of fenoxycarb but this value was reduced to 7.33 ± 1.33 per cent at 0.50 ppm concentration of fenoxycarb. Under same conditions, per cent pupal mortality was found to be 0.69 ±0.70 per cent in control larvae, which was increased to 7.52 ± 1.40 and 100 per cent at 0.001 and 0.50 ppm concentration of fenoxycarb respectively. It is mentionable that at 0.50 ppm concentration of fenoxycarb only 7.33 ± 1.33 per cent pupation occurred but all these pupae get perishe and hence none of the adult emerged at this concentration. A significant reduction in adult emergence was recorded following increased concentrations of this JHA. In control larval group, 95.33 ± 1.23 per cent adult emergence was recorded that decreased to 82.00 ± 0.89 per cent at 0.001 ppm

Table 1: Effect of Fenoxycarb on the Ontogeny of Rice Moth, *Corcyra cephalonica* Exposed as First Instar Larvae

Fenoxycarb Concentration (ppm)	*Fenoxycarb Consumed* (µg/larva)*	*Per cent Larval Mortality#*	*Per cent Pupation#*	*Per cent Pupal Mortality#*	*Per cent Adult Emergence#*	*Per cent Adult Mortality#*	*Per cent Total Mortality#*	*Corrected Total Mortality***
Control	0.00	4.00 ± 1.03	96.00 ± 1.03	0.69 ± 0.70	95.33 ± 1.23	-	4.67 ± 1.23	-
0.001	0.0002	11.33 ± 1.23[b]	88.67 ±1.23[b]	7.52 ± 1.40[b]	82.00 ± 0.89[a]	-	18.00 ± 0.89	13.98 ± 0.94
0.005	0.0013	17.33 ± 2.67[a]	82.67 ± 2.67[a]	8.07 ± 1.43[b]	76.00 ± 1.98[a]	-	24.00 ± 2.07	20.28 ± 2.17
0.01	0.0027	20.00 ± 2.92[a]	80.00 ± 2.92[a]	21.67 ± 2.82[a]	62.67 ± 3.21[a]	7.62 ± 2.18	42.11 ± 3.54	39.27 ± 3.71
0.05	0.0138	31.33 ± 2.40[a]	68.67 ± 2.40[a]	41.75 ± 1.71[a]	40.00 ± 2.30[a]	16.72 ± 3.08	66.67 ± 2.46	65.04 ± 2.95
0.10	0.0284	66.00 ± 2.48[a]	34.00 ± 2.48[a]	43.47 ± 4.13[a]	19.33 ± 2.17[a]	44.44 ± 4.46	88.67 ± 1.90	88.11 ± 2.00
0.50	0.1500	92.67 ± 1.23[a]	7.33 ± 1.23[a]	100[a]	-	-	100	100
1.00	0.3140	100 [a]	-	-	-	-	100	100

Values are expressed as mean ± SEM of six replicates; *Known weight of treated diet was given to each set of 25 larvae. After completion of life-cycle, the remaining food was reweighed to calculate the amount of fenoxycarb consumed per larva (µg/larva) at each concentration of fenoxycarb. a and b Significantly different $p < 0.001$ and $p < 0.01$ respectively compared with control when t- test was applied. Total mortality includes larval mortality, pupal mortality and adult mortality. ** Calculated by Abbott's formula (1925).

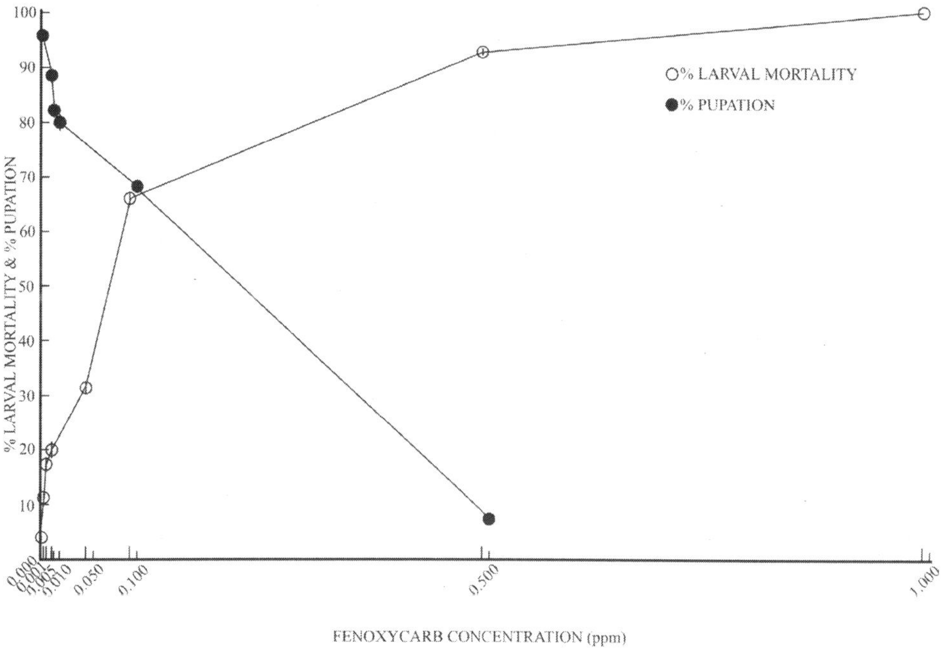

Figure 1: Graphic Representation of Effect of Fenoxycarb on Larval Mortality and Pupation of the Rice Moth, *Corcyra cephalonica* Exposed as First Instar Larvae (Mean ± SEM).

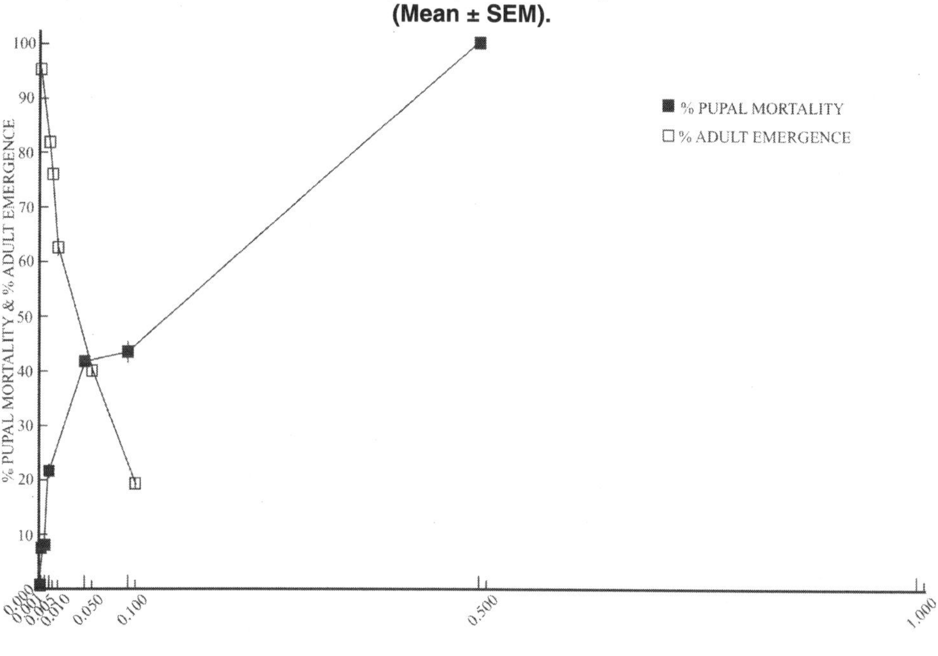

Figure 2: Graphic Representation of Effect of Fenoxycarb on Pupal Mortality and Adult Emergence of the Rice Moth, *Corcyra cephalonica* Exposed as First Instar Larvae (Mean ± SEM).

concentration of fenoxycarb and this value was reduced to be 19.33 ± 2.17 per cent at 0.10 ppm concentration of fenoxycarb.

Toxicity of Fenoxycarb on the Ontogeny of Second Instar Larval Treatment (Table 2, Figures 3 and 4)

The larval mortality was significantly increased with increase in the concentration of fenoxycarb in the diet. In case of control larvae, larval mortality was 2.67 ± 1.33 per cent but at 0.001 ppm concentration of fenoxycarb it was observed to be 4.00 ± 1.46 per cent which increased to 100 per cent at 1.00 ppm concentration of fenoxycarb. As the fenoxycarb concentration increases a significant reduction in pupation and a significant enhancement in pupal mortality occured. In case of control larval groups, 97.33 ± 0.93 per cent pupation was observed that decreased to 94.00 ± 0.89 per cent at 0.001 ppm concentration of fenoxycarb but this value was reduced to 32.00 ± 4.62 per cent at 0.50 ppm concentration of fenoxycarb. Pupal mortality in control larval groups, was found to be 1.37 ± 0.84 per cent which increased to 5.56 ± 0.89 per cent at 0.001 ppm concentration of this compound and 100 per cent at 0.50 ppm concentration of fenoxycarb. There was a significant reduction in adult emergence with increased concentrations of this JHA. In control larval groups, 95.33 ± 1.23 per cent adult emergence was recorded that decreased to 90.67 ± 1.69 per cent and 34.00 ± 2.68 per cent at 0.001 and 0.10 ppm concentrations of fenoxycarb respectively.

Toxicity of Fenoxycarb on the Ontogeny of Third Instar Larval Treatment (Table 3, Figures 5 and 6)

Different concentrations of fenoxycarb caused a significant enhancement in larval and pupal mortality and a significant reduction in pupation and adult emergence. In control larval groups, the larval mortality was 2.00 ± 0.89 per cent, which increased to 4.00 ± 1.46 per cent and 90.00 ± 1.71 per cent following exposure of 0.001 ppm and 0.10 ppm concentrations of fenoxycarb respectively. In case of control larval groups, 98.00 ± 0.89 per cent pupation was recorded. As the fenoxycarb concentration increases an insignificant reduction in pupation up to 0.005 ppm concentration and a significant reduction in pupation beyond that concentration (0.01 to 1.00 ppm) was observed. At 1.00 ppm concentration of fenoxycarb only 10.00 ± 1.71 per cent pupation was recorded. A significant enhancement in pupal mortality occurred with increasing concentrations of fenoxycarb. Pupal mortality in control larval groups, was 2.03 ± 0.91 per cent but at 0.001 ppm concentration of fenoxycarb 4.96 ± 1.76 per cent pupal mortality was recorded that increased to 100 per cent at 0.50 and 1 ppm concentrations of fenoxycarb. A significant reduction in adult emergence was recorded following exposure of increased concentrations of fenoxycarb. In control larval groups, adult emergence was 96.67 ± 1.03 per cent but it was observed to be 91.33 ± 2.81 per cent at 0.001 ppm concentration of fenoxycarb that declined to 46.00 ± 1.71 per cent at 0.10 ppm concentration of fenoxycarb and beyond that there was no adult emergence.

Table 2: Effect of Fenoxycarb on the Ontogeny of Rice Moth, *C. cephalonica* Exposed as Second Instar Larvae

Fenoxycarb Concentration (ppm)	*Fenoxycarb Consumed* (µg/larva)*	*Per cent Larval Mortality#*	*Per cent Pupation#*	*Per cent Pupal Mortality#*	*Per cent Adult Emergence#*	*Per cent Adult Mortality#*	*Per cent Total Mortality#*	*Corrected Total Mortality***
Control	0.0000	2.67 ± 1.33	97.33 ± 0.93	1.37 ± 0.84	95.33 ± 1.23	-	4.00 ± 1.03	-
0.001	0.0002	4.00 ± 1.46	96.00 ± 1.46	5.56 ± 0.89[b]	90.67 ± 1.69[c]	-	9.33 ± 1.69	5.56 ± 1.76
0.005	0.0012	6.00 ± 0.89	94.00 ± 0.89[c]	13.48 ± 1.74[a]	81.33 ± 1.98[a]	-	18.67 ± 1.98	15.28 ± 2.06
0.01	0.0026	14.67 ± 1.98[a]	85.33 ± 1.98[a]	24.99 ± 2.67[a]	64.00 ± 2.30[a]	-	36.00 ± 2.30	33.33 ± 2.41
0.05	0.0133	22.00 ± 2.00[a]	78.00 ± 1.98[a]	27.35 ± 2.81[a]	56.67 ± 2.81[a]	17.40 ± 2.62	53.19 ± 2.23	51.23 ± 2.32
0.10	0.0278	43.33 ± 3.78[a]	56.67 ± 3.78[a]	40.00 ± 3.38[a]	34.00 ± 2.68[a]	21.44 ± 4.25	73.29 ± 2.46	72.17 ± 2.56
0.50	0.1450	68.00 ± 4.62[a]	32.00 ± 4.62[a]	100[a]	-	-	100	100
1.00	0.3080	100[a]	-	-	-	-	100	100

Values are expressed as mean ± SEM of six replicates. *Known weight of treated diet was given to each set of 25 larvae. After completion of life-cycle, the remaining food was reweighed to calculate the amount of fenoxycarb consumed per larva (µg/larva) at each concentration of fenoxycarb. a, b and c Significantly different $p < 0.001$, $p < 0.01$ and $p < 0.05$, respectively compared with control when t- test was applied. Total mortality includes larval mortality, pupal mortality and adult mortality. ** Calculated by Abbott's formula (1925).

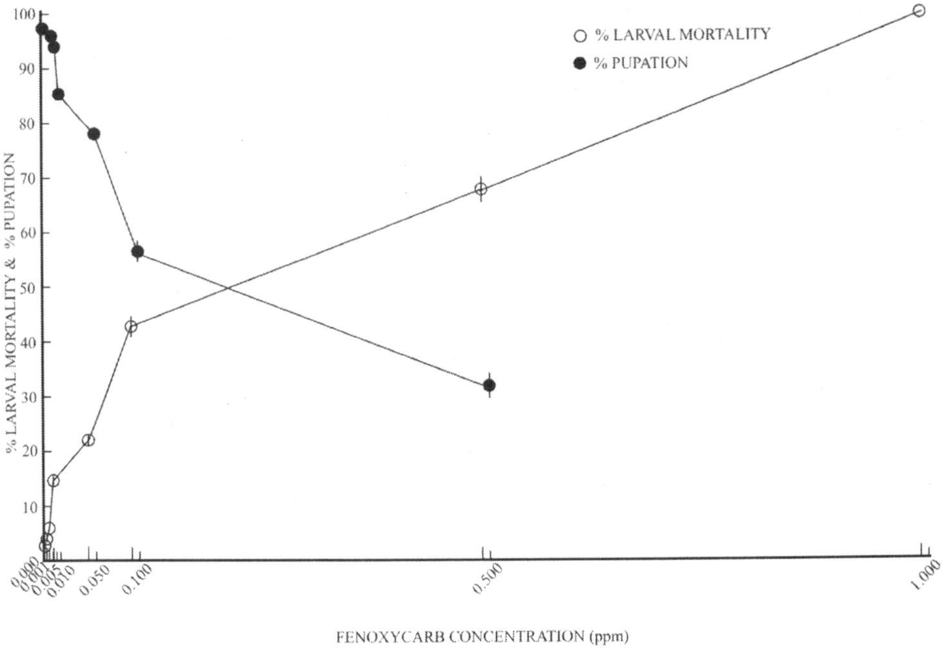

Figure 3: Graphic Representation of Effect of Fenoxycarb on Larval Mortality and Pupation of the Rice Moth, *Corcyra cephalonica* Exposed as Second Instar Larvae (Mean ± SEM).

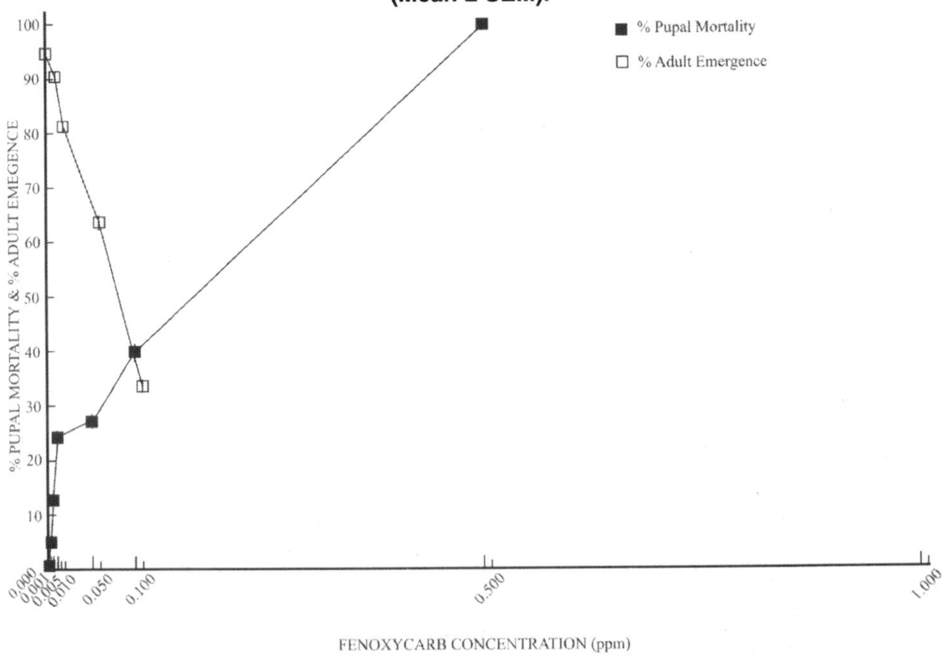

Figure 4: Graphic Representation of Effect of Fenoxycarb on Pupal Mortality and Adult Emergence of the Rice Moth, *Corcyra cephalonica* Exposed as Second Instar Larvae (Mean ± SEM).

Table 3: Effect of Fenoxycarb on the Ontogeny of Rice Moth, *C. cephalonica* Exposed as Third Instar Larvae

Fenoxycarb Concentration (ppm)	*Fenoxycarb Consumed* (μg/larva)*	*Per cent Larval Mortality#*	*Per cent Pupation#*	*Per cent Pupal Mortality#*	*Per cent Adult Emergence#*	*Per cent Adult Mortality#*	*Per cent Total Mortality#*	*Corrected Total Mortality***
Control	0.00	2.00 ± 0.89	98.00 ± 0.89	2.03 ± 0.91	96.00 ± 1.03	-	4.00 ± 1.03	-
0.001	0.0002	4.00 ± 1.46	96.00 ± 1.46	4.96 ± 1.76	91.33 ± 2.81	-	8.67 ± 2.81	4.86 ± 2.93
0.005	0.0012	5.33 ± 1.33	94.67 ± 1.33	9.16 ± 1.28[b]	86.00 ± 1.71[a]	-	14.00 ± 1.71	10.42 ± 1.78
0.01	0.0025	7.33 ± 1.23[b]	92.67 ± 1.23[b]	13.67 ±1.85[a]	80.00 ± 2.53[a]	-	20.00 ± 2.53	16.67 ± 2.63
0.05	0.0130	11.33 ± 1.91[b]	88.67 ± 1.91[b]	11.90 ± 2.13[b]	78.00 ± 1.71[a]	21.35 ±1.10	38.67 ± 1.97	36.11 ± 2.06
0.10	0.0272	46.00 ± 1.71[a]	54.00 ± 1.71[a]	14.80 ± 1.83[a]	46.00 ± 1.71[a]	31.97 ± 2.94	68.67 ±1.91	67.36 ± 1.98
0.50	0.1420	60.00 ± 4.26[a]	40.00 ± 4.26[a]	100[a]	-	-	100	100
1.00	0.2960	90.00 ± 1.71[a]	10.00 ± 1.71[a]	100[a]	-	-	100	100

Values are expressed as mean ± SEM of six replicates. *Known weight of treated diet was given to each set of 25 larvae. After completion of life-cycle, the remaining food was reweighed to calculate the amount of fenoxycarb consumed per larva (μg/larva) at each concentration of fenoxycarb. a and b Significantly different $p < 0.001$ and $p < 0.01$ respectively compared with control when t- test was applied. Total mortality includes larval mortality, pupal mortality and adult mortality. **Calculated by Abbott's formula (1925).

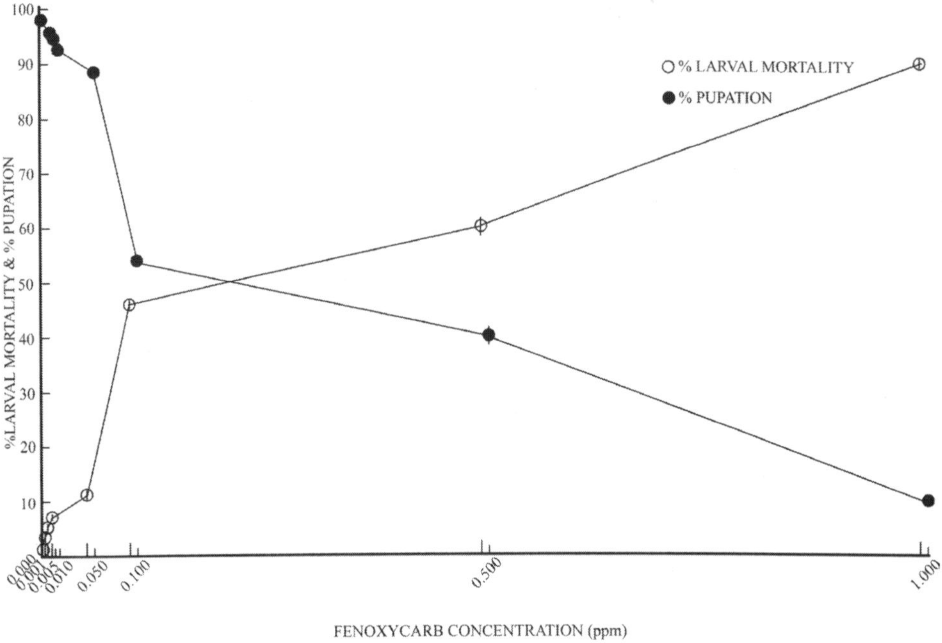

Figure 5: Graphic Representation of Effect of Fenoxycarb on Larval Mortality and Pupation of the Rice Moth, *Corcyra cephalonica* Exposed as Third Instar Larvae (Mean ± SEM).

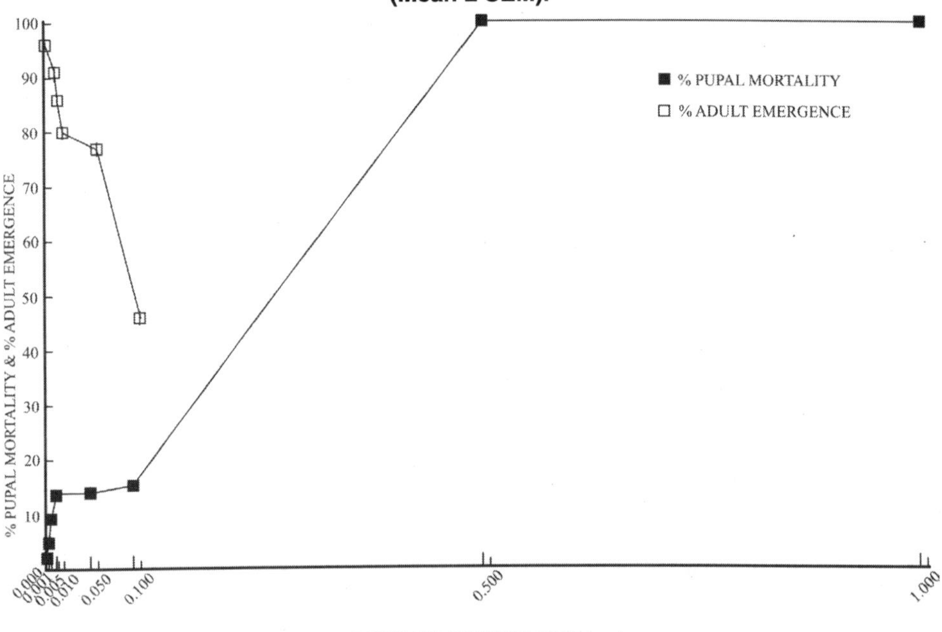

Figure 6: Graphic Representation of Effect of Fenoxycarb on Pupal Mortality and Adult Emergence of the Rice Moth, *Corcyra cephalonica* Exposed as Third Instar Larvae (Mean ± SEM).

Toxicity of Fenoxycarb on the Ontogeny of Fourth Instar Larval Treatment (Table 4, Figures 7 and 8)

A significant larval mortality was obtained with the increase of fenoxycarb concentration in the diet. In case of control larvae, larval mortality was recorded to be 1.33 ± 0.84 per cent but at 0.001 ppm concentration of fenoxycarb larval mortality was observed to be 2.67 ± 0.84 per cent which increased to 85.00 ± 1.71 per cent at 1.00 ppm concentration of fenoxycarb. As the fenoxycarb concentration increases a significant reduction in pupation and a significant enhancement in pupal mortality did occur. In case of control larval groups, 98.67 ± 0.84 per cent pupation was observed that decreased to 97.33 ± 0.84 per cent at 0.001 ppm concentration of fenoxycarb but this value was reduced to 15.00 ± 1.71 per cent at 1.00 ppm concentration of fenoxycarb. Under same conditions, per cent pupal mortality was found to be 2.03 ± 0.89 per cent in control larvae, which was increased to 6.84 ± 0.92 per cent at 0.001 ppm and 100 per cent at 0.50 and 1.00 ppm concentrations of fenoxycarb. It is mentionable that at 0.50 and 1.00 ppm concentrations of fenoxycarb pupation took place but all the pupae get perished and hence none of the adults emerged at these two concentrations of fenoxycarb. A significant reduction in adult emergence was recorded following increased concentrations of this JHA. In control larval groups, 96.67 ± 0.67 per cent adult emergence was recorded that decreased to 90.67 ± 1.69 per cent at 0.001 ppm concentration of fenoxycarb and this value was reduced to 40.00 ± 2.07 per cent at 0.10 ppm concentration of fenoxycarb.

Along with above mentioned facts we also found that there were delayed emergences of adults as the concentrations of fenoxycarb increased in diet. At 0.01, 0.05 and 0.10 ppm concentrations of fenoxycarb (in first instar larvae) many of emerged adults were abnormal. Degree of abnormalities ranges from no morphological distinguishing clue for male and female, unfolded or twisted wings, twisted legs and abnormally long abdomen in males and too much swollen abdomen in females. Majority of abnormal adults were died within 24 h of their emergence. However, the normal and quite healthy adults were also emerged along with abnormal ones at all concentrations where adult emergence occurred but their percentages were decreased with increased concentrations.

It is noteworthy that first instar larvae exposed to different concentrations of fenoxycarb revealed maximum toxicity in comparision to second, third and fourth instar larvae exposed to such concentrations of fenoxycarb. Their order of toxicity decreased in second, third and fourth instar treated larvae. It may be accounted that this difference in the toxicity of fenoxycarb was due to difference in age of larval treatment and tenure of exposure of the JHA. In first instar larval treatment the tenure of exposure was maximum, which revealed maximum toxicity to the insect but in consecutive instars the tenure of exposure gradually decreased that resulted in decreased order of toxicity in second, third and fourth instars. The fourth instar larvae exposed to fenoxycarb revealed comparatively poor toxicity and this is ofcourse due to reduced tenure of exposure of fenoxycarb to the fourth instar larvae.

LD_{50} Values (μg/larva), 95 per cent confidence limits (lower and upper confidence limits) of LD_{50} slope values and heterogeneity of fenoxycarb to the first

Table 4: Effect of Fenoxycarb on the Ontogeny of Rice Moth, *C. cephalonica* Exposed as Fourth Instar Larvae

Fenoxycarb Concentration (ppm)	*Fenoxycarb Consumed* (µg/larva)*	*Per cent Larval Mortality#*	*Per cent Pupation#*	*Per cent Pupal Mortality#*	*Per cent Adult Emergence#*	*Per cent Adult Mortality#*	*Per cent Total Mortality#*	*Corrected Total Mortality***
Control	0.0000	1.33 ± 0.84	98.67 ± 0.84	2.03 ± 0.89	96.67 ± 0.67	-	3.33 ± 0.67	-
0.001	0.0001	2.67 ± 0.84	97.33 ± 0.84	6.84 ± 0.92[b]	90.67 ± 1.69[b]	-	8.67 ± 1.23	5.52 ± 1.35
0.005	0.0804	4.67 ± 1.23[c]	95.33 ± 1.23[c]	8.39 ± 1.29[b]	87.33 ± 1.90[b]	-	12.00 ± 1.79	8.97 ± 1.85
0.01	0.0181	17.33 ± 1.98[a]	82.67 ± 1.98[a]	21.77 ± 3.66[a]	64.67 ± 3.49[a]	-	35.33 ± 3.49	33.11 ± 3.61
0.05	0.0102	21.33 ± 2.23[a]	78.67 ± 2.23[a]	26.27 ± 2.15[a]	58.00 ± 2.68[a]	14.93 ± 0.74	50.67 ± 3.68	48.97 ± 2.31
0.10	0.0214	32.67 ± 3.17[a]	67.33 ± 3.17[a]	42.59 ± 2.63[a]	40.00 ± 2.07[a]	21.24 ± 2.32	68.40 ± 1.23	67.31 ± 1.77
0.50	0.1160	53.33 ± 2.67[a]	46.67 ± 2.67[a]	100[a]	-	-	100	100
1.00	0.2560	85.00 ± 1.71[a]	15.00 ± 1.71[a]	100[a]	-	-	100	100

Values are expressed as mean ± SEM of six replicates. *Known weight of treated diet was given to each set of 25 larvae. After completion of life-cycle, the remaining food was reweighed to calculate the amount of fenoxycarb consumed per larva (µg/larva) at each concentration of fenoxycarb. a, b and c Significantly different $p < 0.001$, $p < 0.01$ and $p < 0.05$ respectively compared with control when t- test was applied. Total mortality includes larval mortality, pupal mortality and adult mortality. ** Calculated by Abbott's formula (1925).

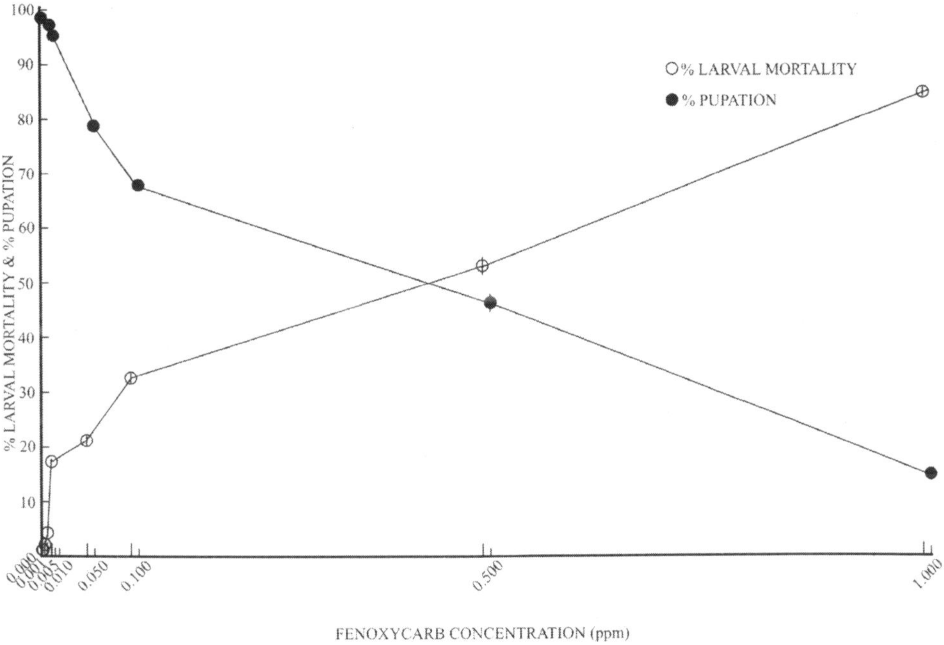

Figure 7: Graphic Representation of Effect of Fenoxycarb on Larval Mortality and Pupation of the Rice Moth, *Corcyra cephalonica* Exposed as Fourth Instar Larvae (Mean ± SEM).

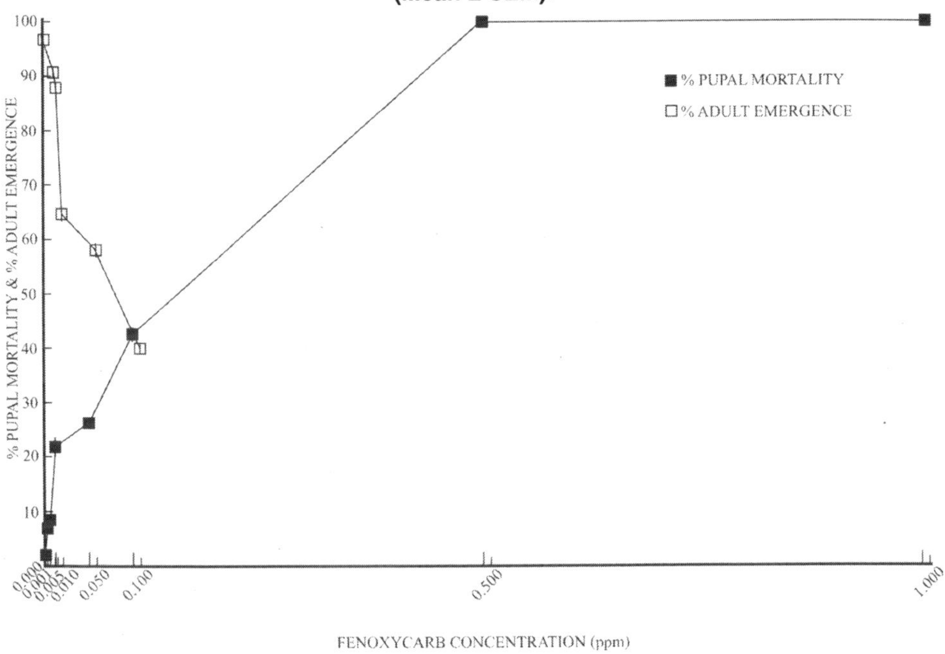

Figure 8: Graphic Representation of Effect of Fenoxycarb on Pupal Mortality and Adult Emergence of the Rice Moth, *Corcyra cephalonica* Exposed as Fourth Instar Larvae (Mean ± SEM).

to fourth instar larvae have been given in Table 5. LD_{50} value was calculated against the insecticide consumed by larvae (µg/larva) and not against the ppm concentration of fenoxycarb in food. It is evident from the table that fenoxycarb was most toxic against the first instar larvae and least toxic against fourth instar larvae.

Table 5: LD_{10}, LD_{50} and LD_{90} Values, Confidence Limits (LCL and UCL) of LD_{10}, LD_{50} and LD_{90}, Slope Values and Heterogeneity of Fenoxycarb to the I-IV Instar Larvae of Rice Moth, *C. cephalonica*

Insecticide	*Instar*	*Effective Dose (µg/larva)*	*95 per cent Confidence Limits of LD_{50} LCL UCL*	*Slope Values*	*Heterogeneity*
Fenoxycarb	I	LD_{10} 0.007	0.006 0.009	2.362 ± 0.145	0.94
		LD_{50} 0.026	0.023 0.029		
		LD_{90} 0.090	0.074 0.115		
Fenoxycarb	II	LD_{10} 0.010	0.008 0.014	1.903 ± 0.103	1.94
		LD_{50} 0.049	0.040 0.061		
		LD_{90} 2.33	0.172 0.344		
Fenoxycarb	III	LD_{10} 0.012	0.009 0.015	1.728 ± 0.092	1.70
		LD_{50} 0.067	0.055 0.083		
		LD_{90} 0.396	0.268 0.555		
Fenoxycarb	IV	LD_{10} 0.009	0.006 0.012	1.455 ± 0.089	1.67
		LD_{50} 0.068	0.054 0.088		
		LD_{90} 0.518	0.343 0.895		

LCL = Lower Confidence Limit; UCL = Upper Confidence Limit.

SECTION I B

Effect of Fenoxycarb on the Growth Duration and Logevity of Adults of *C. cephalonica*

The growth duration (time taken from egg laying to adult emergence) and longevity (life-span) of adult males and females, were observed in days when I, II, III and IV instar larvae were exposed to different concentrations of fenoxycarb in the rearing medium (Tables 6–9, Figures 9–16). It is noteworthy that females had slightly prolonged growth duration than males in control.

Effect of Fenoxycarb on the Growth Duration of First Instar Larval Treatment (Table 6 and Figure 9)

It was observed that growth duration for the control moth was 37.67 ± 0.67 days for males and 40.83 ± 0.95 days for females. With the increase of fenoxycarb concentration, in the rearing medium, the growth duration was found to be significantly increased for both the sexes. In case males, the growth duration was recorded to be 41.33 ± 0.67 days at 0.001 ppm concentration of fenoxycarb that was found to be enhanced 86.50 ± 1.65 days at 0.10 ppm concentration of this JHA. The growth duration of females was observed to be increased 43.83 ± 1.01 days at 0.001 ppm concentration which was further enhanced to 87.33 ± 1.80 days at 0.10 ppm concentration of fenoxycarb.

Table 6: Effect of Fenoxycarb on Growth Duration and Longevity of Adults of Rice Moth, *C. cephalonica* Exposed as First Instar Larvae

Fenoxycarb Concentration (ppm)	*Growth Duration# of Males (Days)*	*Growth Duration# of Females (Days)*	*Adult Longevity of Males (Days)*	*Adult Longevity# of Females (Days)*
Control	37.67 ± 0.67	40.83 ± 0.95	13.50 ± 0.42	8.50 ± 0.43
0.001	41.33 ± 0.67[b]	43.83 ± 1.01	11.50 ± 0.62[c]	7.67 ± 0.56
0.005	44.67 ± 1.52[b]	48.17 ± 1.51[b]	9.33 ± 0.42[a]	7.00 ± 0.68
0.01	53.17 ± 1.60[a]	54.83 ± 1.35[a]	7.00 ± 1.95[b]	4.67 ± 1.23[c]
0.05	80.00 ± 2.14[a]	82.83 ± 1.54[a]	6.17 ± 1.66[b]	4.33 ± 1.09[b]
0.10	86.50 ± 1.65[a]	87.33 ± 1.80[a]	5.00 ± 0.86[a]	4.00 ± 0.97[b]
0.50	92.50 ± 1.88[a*]	92.50 ± 1.88[a*]	-	-
1.00	105.67 ± 3.51[a*]	105.67 ± 3.51[a*]	-	-

Values are expressed as mean ± SEM of six replicates. a, b, c and d significantly different $p < 0.001$, $p < 0.01$ and $p < 0.05$ respectively compared with control when t- test was applied.

Straight line regression equation was applied between different concentrations of fenoxycarb and their corresponding growth duration of males and females longevity of adults to observe the significant correlation:

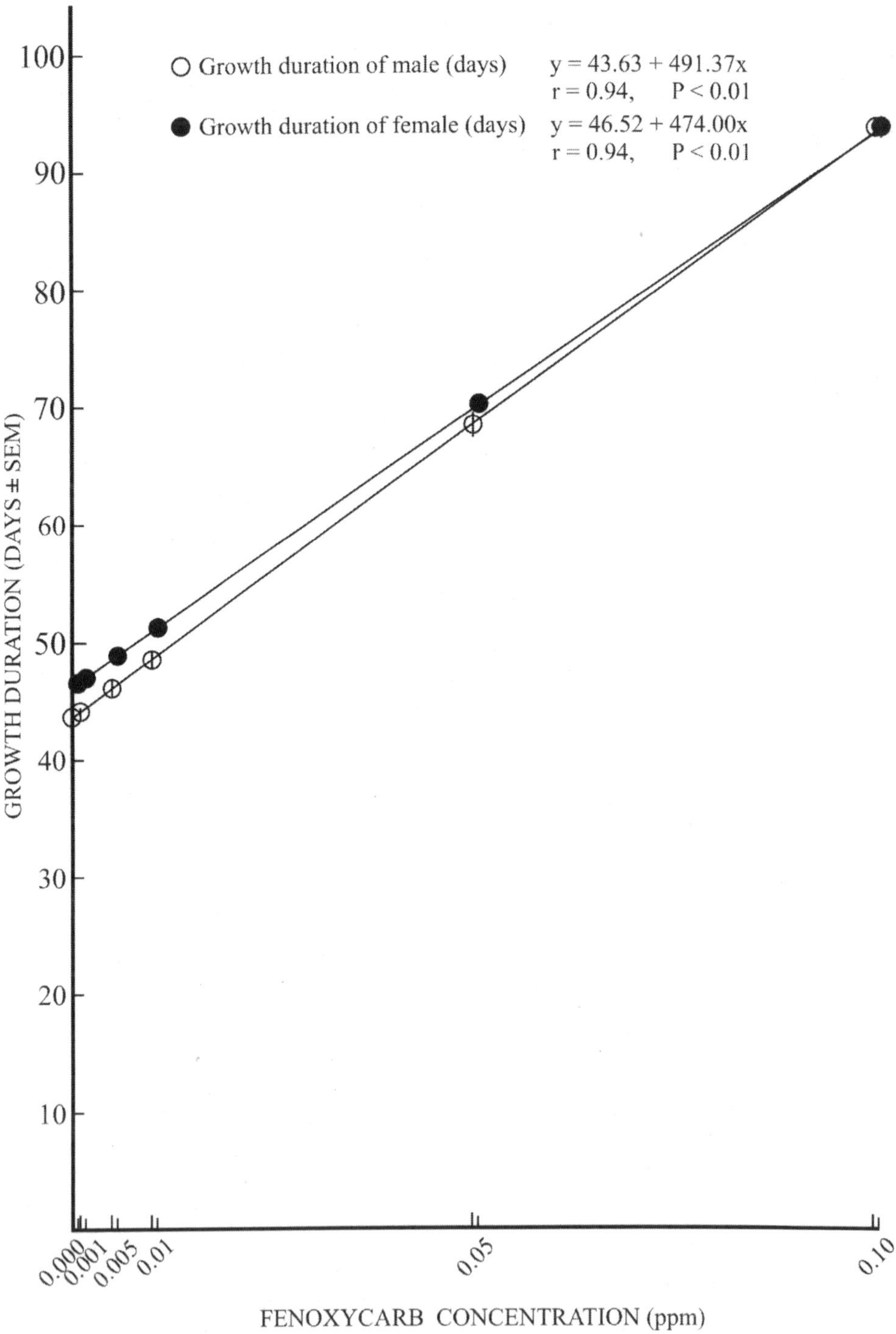

Figure 9: Graphic Representation of Effect of Fenoxycarb on the Growth Duration of the Rice Moth, *Corcyra cephalonica* Exposed as First Instar Larvae (Mean ± SEM).

Growth duration of males	y = 43.63 + 491.37x;	r = 0.94	p < 0.01
Growth duration of females	y = 46.52 + 474.00x;	r = 0.94	p < 0.01
Longevity of male adults	y = 10.53 - 64.21x;	r = - 0.78	p insignificant
Longevity of female adults	y = 6.06 - 0.41x;	r = - 0.76	p insignificant

* Larval tenure since adults did not emerge, so these concentrations have not been accounted in correlation.

It is noteworthy that at 0.50 and 1.00 ppm concentrations of fenoxycarb adults did not emerge but larval period was much prolonged wich lead to the production of giant larvae and supernumerary larvae, and their average life-span was recorded to be 92.50 ± 1.88 and 105.67 ± 3.51 days respectively.

Effect of Fenoxycarb on the Logevity of Adults of First Instar Larval Treatment (Table 6 and Figure 10)

The life-span of control moth was recorded to be 13.50 ± 0.42 days for males and 8.50 ± 0.43 days for females. With the increased concentration of this JHA the longevity of male and female individuals was found to be significantly reduced. At 0.001 ppm concentration of fenoxycarb male adults longevity was 11.50 ± 0.62 days while female adults longevity was 7.67 ± 0.56 days. But male and female adults longevity was observed to be 5.00 ± 0.86 and 4.00 ± 0.97 days respectively at 0.10 ppm concentration of fenoxycarb. It deserves mention that at 0.01, 0.05 and 0.10 ppm concentrations of fenoxycarb many of the emerged adults were died within 24 h of their emergence.

Effect of Fenoxycarb on the Growth Duration of Second Instar Larval Treatment (Table 7 and Figure 11)

In case of second instar larval treatment the growth duration of control male moth was found to be 37.83 ± 0.70 days and growth duration of female moth was 40.83 ± 0.87 days which significantly increased with increase in concentrations of fenoxycarb. At 0.001 ppm concentration the growth duration of male and female was 41.33 ± 0.67 and 43.83 ± 1.01 days respectively which increased to 85.17 ± 1.81 and 86.50 ± 1.98 days for males and females respectively at 0.10 ppm concentration of fenoxycarb. It is also notable that at 0.50 and 1.00 ppm concentrations of fenoxycarb larval period was much prolonged which lead to production of giant larvae, supernumerary larvae and their average life-span was recorded to be 90.17 ± 2.41 and 103.67 ± 2.95 days respectively. These larvae did not emerged as adults so whether these were male or female could not be determined.

Effect of Fenoxycarb on the Logevity of Adults of Second Instar Larval Treatment (Table 7 and Figure 12)

In case of control males, the adult longevity was found to be 13.67 ± 0.33 days, which significantly reduced to 11.50 ± 0.43 days at 0.001 ppm and 5.50 ± 1.48 days at 0.10 ppm concentrations of fenoxycarb respectively.

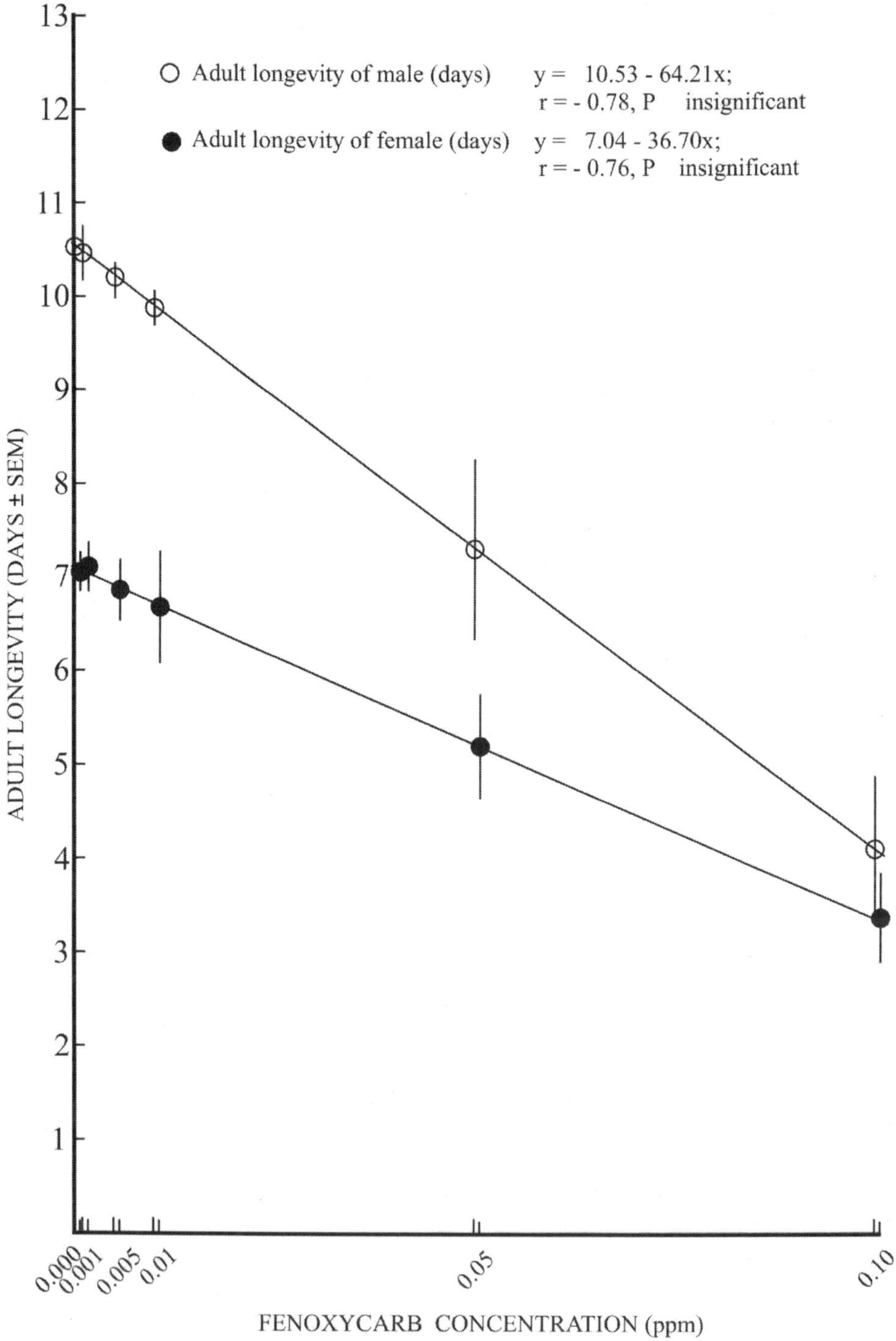

Figure 10: Graphic Representation of Effect of Fenoxycarb on the Adult Longevity of the Rice Moth, *Corcyra cephalonica* Exposed as First Instar Larvae (Mean ± SEM).

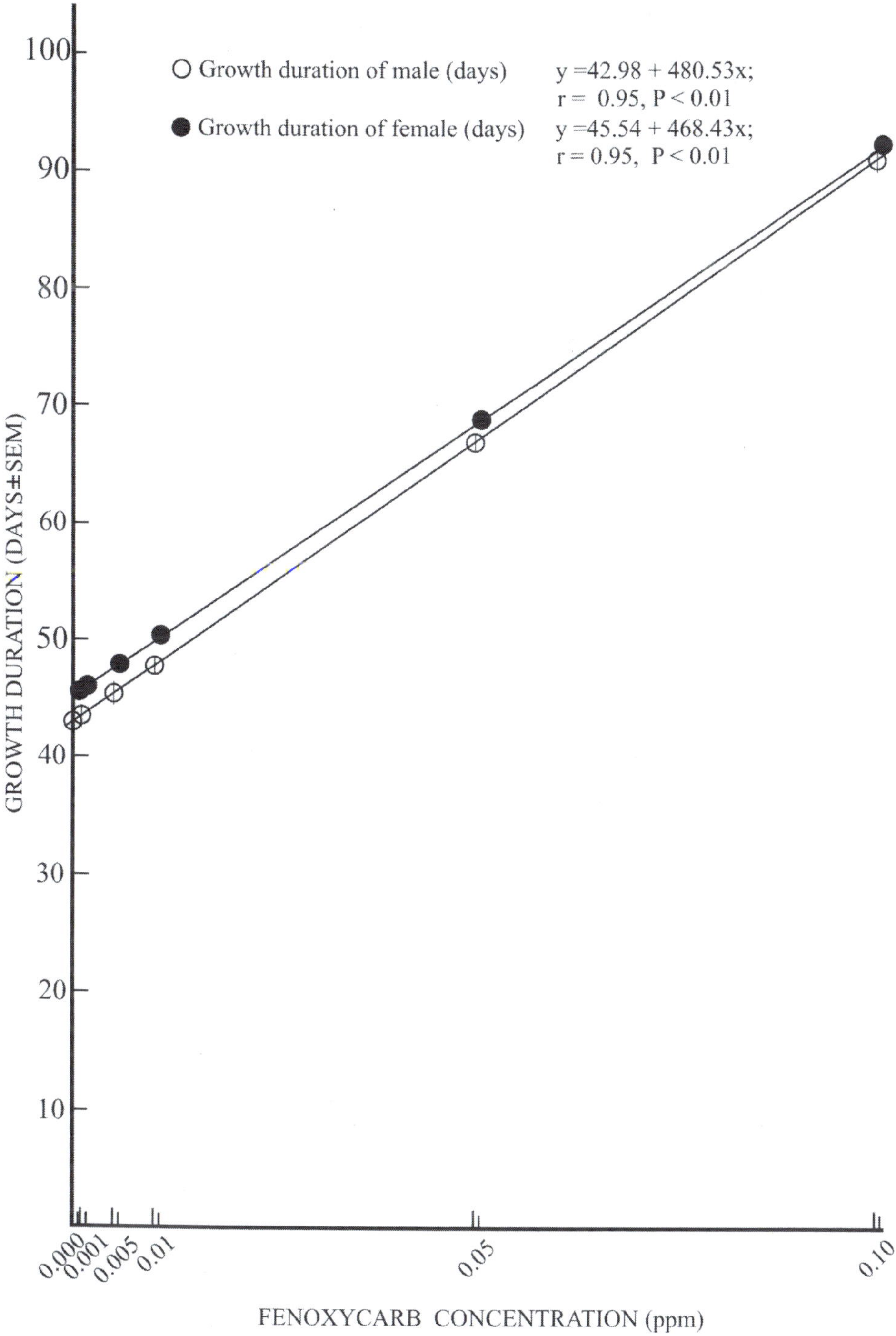

Figure 11: Graphic Representation of Effect of Fenoxycarb on the Growth Duration of the Rice Moth, *Corcyra cephalonica* Exposed as Second Instar Larvae (Mean ± SEM).

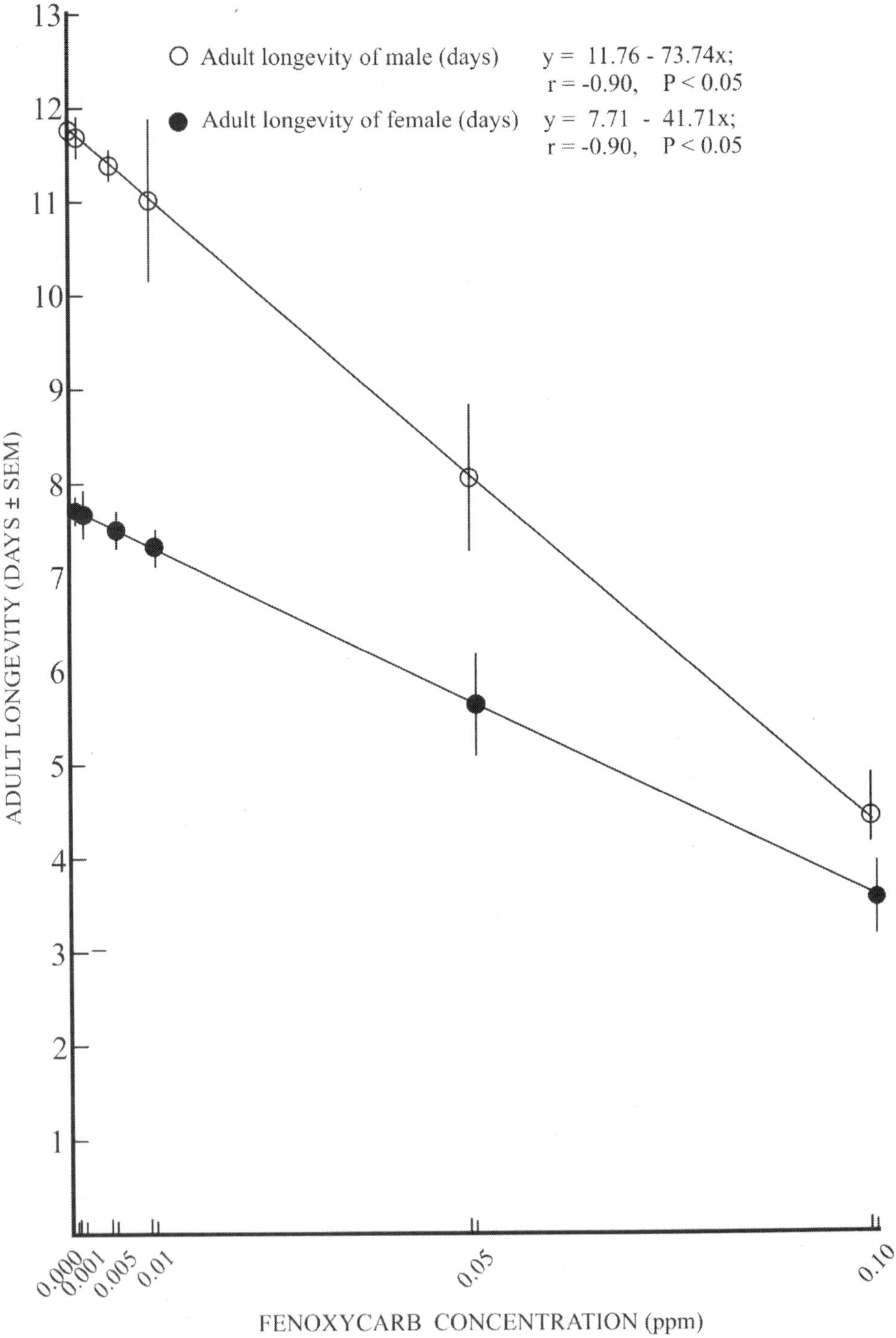

Figure 12: Graphic Representation of Effect of Fenoxycarb on the Adult Longevity of the Rice Moth, *Corcyra cephalonica* Exposed as Second Instar Larvae (Mean ± SEM).

Table 7: Effect of Fenoxycarb on Growth Duration and Longevity of Adults of Rice Moth, *C. cephalonica* Exposed as Second Instar Larvae

Fenoxycarb Concentration (ppm)	*Growth Duration# of Males (Days)*	*Growth Duration# of Females (Days)*	*Adult Longevity of Males (Days)*	*Adult Longevity# of Females (Days)*
Control	37.83 ± 0.70	40.83 ± 0.87	13.67 ± 0.33	8.67 ± 0.33
0.001	41.17 ± 0.87[c]	42.83 ± 0.60	11.50 ± 0.43[b]	8.00 ± 0.52
0.005	43.50 ± 1.52[b]	46.83 ± 1.60[b]	11.33 ± 0.49[b]	7.33 ± 0.49[c]
0.01	51.83 ± 1.58[a]	53.83 ± 1.14[a]	10.33 ± 0.49[a]	6.67 ± 0.56[c]
0.05	78.17 ± 1.36[a]	80.17 ± 0.98[a]	6.00 ± 1.67[b]	4.50 ± 1.15[b]
0.10	85.17 ± 1.81[a]	86.50 ± 1.98[a]	5.50 ± 1.48[a]	4.17 ± 1.05[b]
0.50	90.17 ± 2.41[a*]	90.17 ± 2.41[a*]	-	-
1.00	103.67 ± 2.95[a*]	103.67 ± 2.95[a*]	-	-

Values are expressed as mean ± SEM of six replicates. a, b, c and d significantly different $p < 0.001$, $p < 0.01$ and $p < 0.05$ respectively compared control when t- test was applied.

Straight line regression equation was applied between different concentrations of fenoxycarb and their corresponding growth duration of males and females and longevity of adults to observe the significant correlation:

Growth duration of males	$y = 42.98 + 480.53x$;	$r = 0.95$	$p < 0.01$
Growth duration of females	$y = 45.54 + 468.43x$;	$r = 0.95$	$p < 0.01$
Longevity of male adults	$y = 11.76 - 73.74x$;	$r = - 0.90$	$p < 0.05$
Longevity of female adults	$y = 7.71 - 41.71x$;	$r = - 0.90$	$p < 0.05$

* Larval tenure since adults did not emerge, so these concentrations have not been accounted in correlation.

The longevity of control females, was recorded as 8.67 ± 0.33 days, which also reduced with increasing concentrations of fenoxycarb. At 0.001 ppm concentration of fenoxycarb the longevity of female was 8.00 ± 0.52 days which reduced to 4.17 ± 1.05 days at 0.10 ppm concentration of fenoxycarb. In addition at 0.05 and 0.10 ppm concentrations of fenoxycarb many of emerged adults were died within 24 h of their emergence.

Effect of Fenoxycarb on the Growth Duration of Third Instar Larval Treatment (Table 8 and Figure 13)

In the third intar larval treatment, the growth duration of control moth was recorded 37.17 ± 0.48 days for males and 40.33 ± 0.49 days for females. The growth duration was found to be increased with increase in the concentrations of fenoxycarb in rearing medium. The growth duration of males and females were 37.33 ± 0.42 and 41.17 ± 0.48 days respectively at 0.001 ppm concentration of fenoxycarb which further increased to 85.17 ± 1.85 and 86.00 ± 2.89 day for males and females respectively at 0.10 ppm concentrations of fenoxycarb. It deserves mention that at 0.50 and 1.00 ppm concentrations of fenoxycarb adults (neither males nor females) did not emerge

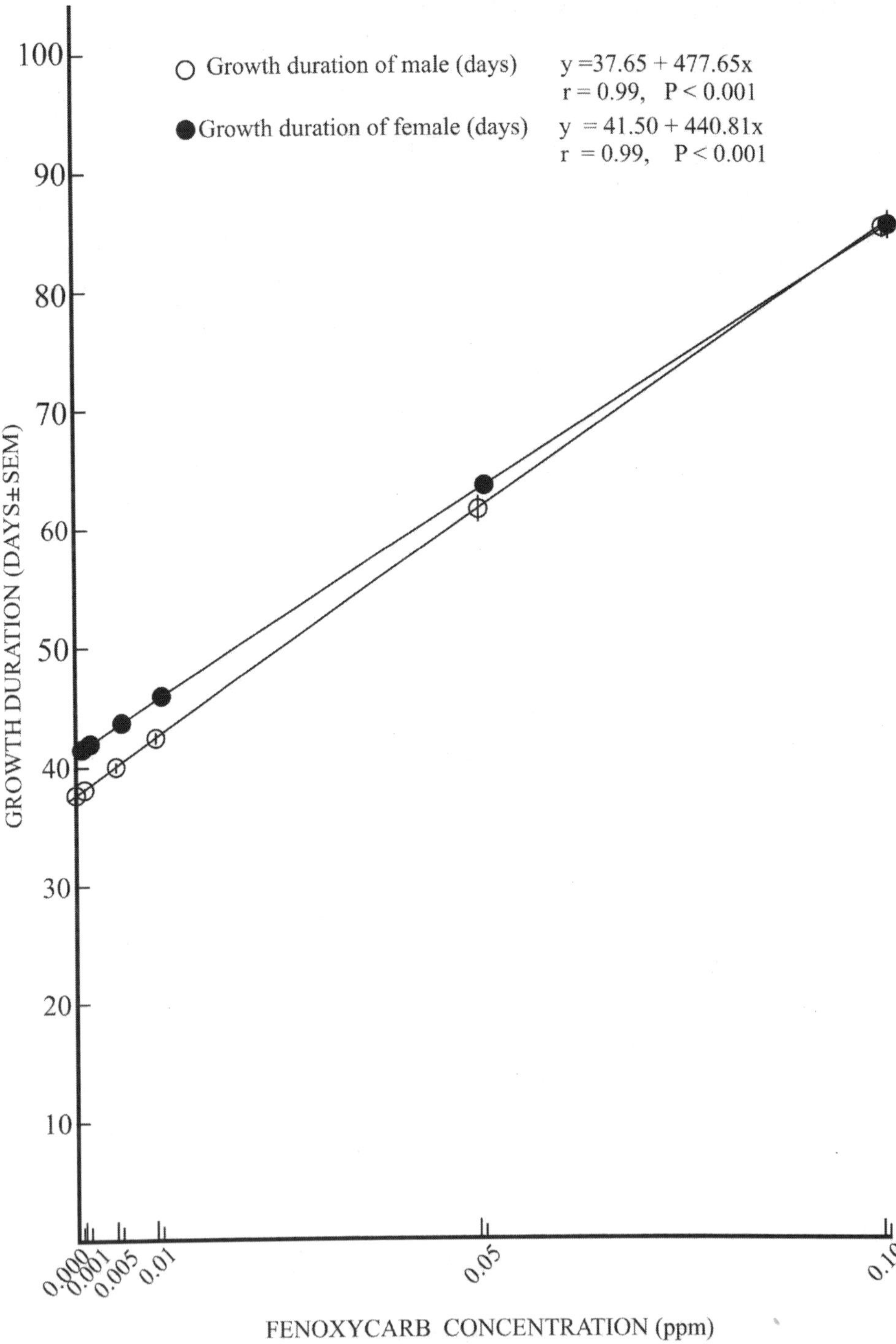

Figure 13: Graphic Representation of Effect of Fenoxycarb on the Growth Duration of the Rice Moth, *Corcyra cephalonica* Exposed as Third Instar Larvae (Mean ± SEM).

but the larval period was observed to be much prolonged which lead to production of giant larvae, supernumerary larvae and their average life-span was recorded to be 90.00 ± 1.80 days and 102 ± 2.66 days respectively.

Table 8: Effect of Fenoxycarb on Growth Duration and Longevity of Adults of Rice Moth, *C. cephalonica* Exposed as Third Instar Larvae

Fenoxycarb Concentration (ppm)	*Growth Duration# of Males (Days)*	*Growth Duration# of Females (Days)*	*Adult Longevity of Males (Days)*	*Adult Longevity# of Females (Days)*
Control	37.17 ± 0.48	40.33 ± 0.49	13.50 ± 0.43	8.50 ± 0.42
0.001	37.33 ± 0.42	41.17 ± 0.48	13.67 ± 0.67	7.50 ± 1.18
0.005	41.83 ± 0.83[a]	46.67 ± 0.92[a]	11.67 ± 1.98	7.83 ± 0.60
0.01	41.67 ± 0.89[b]	45.50 ± 0.99[a]	11.17 ± 1.20	7.17 ± 0.75
0.05	62.00 ± 2.27[a]	62.50 ± 1.31[a]	8.67 ± 2.29	5.50 ± 1.91
0.10	85.17 ± 1.85[a]	86.00 ± 2.89[a]	8.17 ± 2.15[c]	4.89 ± 1.74
0.50	90.00 ± 1.80[a*]	90.00 ± 1.80[a*]	-	-
1.00	102.00 ± 2.66[a*]	102.00 ± 2.66[a*]	-	-

Values are expressed as mean ± *SEM* of six replicates. a, b and c significantly different $p < 0.001$, $p < 0.01$ and $p < 0.05$ respectively compare with control when t- test was applied.

Straight line regression equation was applied between different concentrations of fenoxycarb and their corresponding growth duration of males and females and longevity of adults to observe the significant correlation:

Growth duration of males	$y = 37.65 + 477.65x$;	$r = 0.99$	$p < 0.001$
Growth duration of females	$y = 41.50 + 440.81x$;	$r = 0.99$	$p < 0.001$
Longevity of male adults	$y = 12.57 - 51.70x$;	$r = -0.89$	$p < 0.05$
Longevity of female adults	$y = 7.80 - 32.71x$;	$r = -0.93$	$p < 0.01$

* Larval tenure since adults did not emerge, so these concentrations have not been accounted in correlation.

Effect of Fenoxycarb on the Logevity of Adults of Third Instar Larval Treatment (Table 8 and Figure 14)

When the third instar larvae were exposed to different concentrations of fenoxycarb in the rearing medium, the life-span of the adults, emerged from these larvae, were found to be reduced. The life span of control moth was recorded to be 13.50 ± 0.43 days for males and 8.50 ± 0.42 days for females. The longevity of males were found to be similar to control at 0.001 ppm *i.e.* 13.67 ± 0.67 days which significantly reduced to 8.17 ± 2.15 days at 0.10 ppm concentration of fenoxycarb.

The longevity of females were found to be reduced to 7.50 ± 1.18 days at 0.001 and 4.89 ± 1.74 days at 0.10 ppm concentrations of fenoxycarb. At 0.05 and 0.10 ppm concentrations of fenoxycarb many of emerged adults were died within 24 h of their emergence.

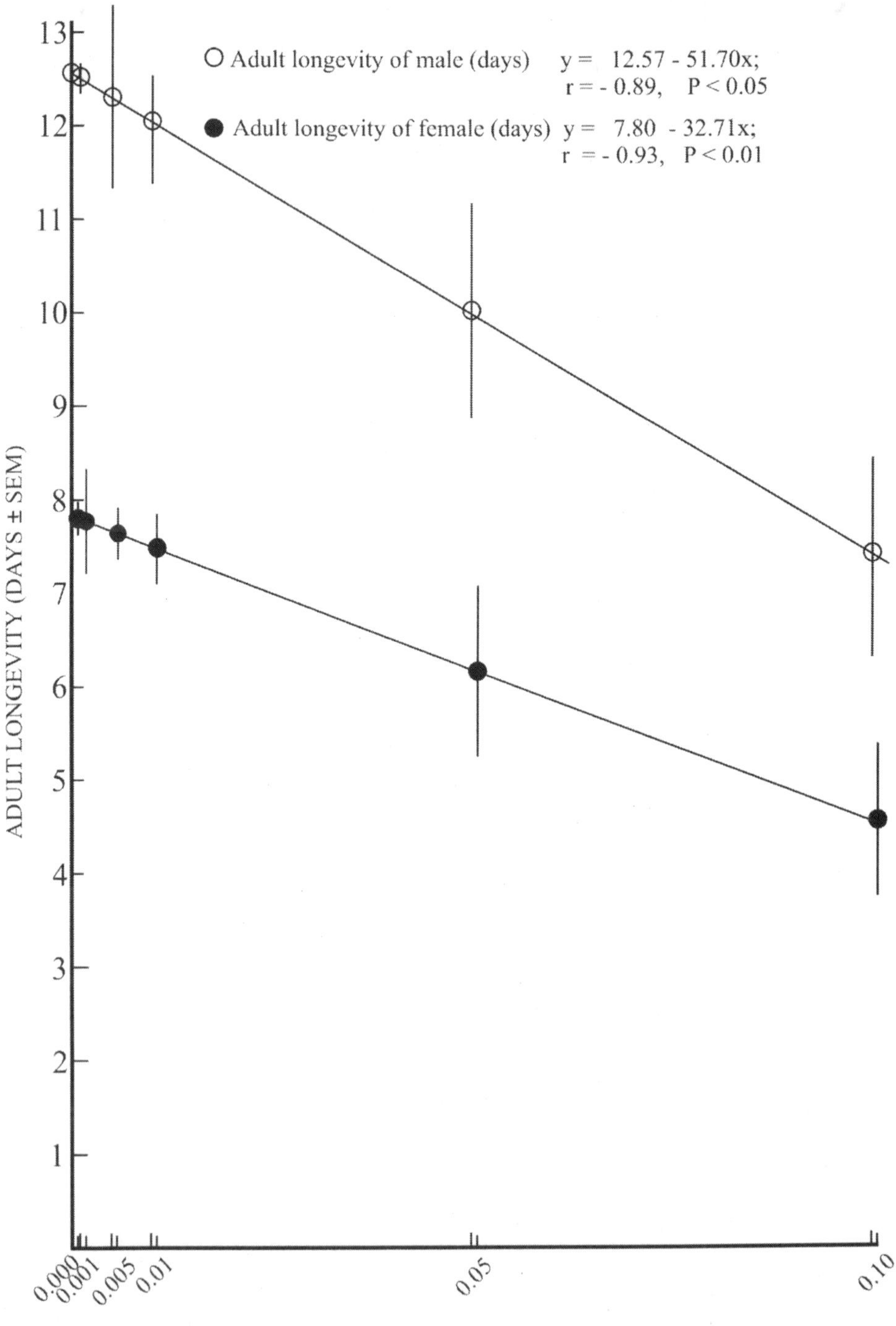

Figure 14: Graphic Representation of Effect of Fenoxycarb on the Adult Longevity of the Rice Moth, *Corcyra cephalonica* Exposed as Third Instar Larvae (Mean ± SEM).

Table 9: Effect of Fenoxycarb on Growth Duration and Longevity of Adults of Rice Moth, *C. cephalonica* Exposed as Fourth Instar Treated Larvae

Fenoxycarb Concentration (ppm)	*Growth Duration# of Males (Days)*	*Growth Duration# of Females (Days)*	*Adult Longevity of Males (Days)*	*Adult Longevity# of Females (Days)*
Control	37.83 ± 0.48	41.83 ± 1.19	13.50 ± 0.43	8.83 ± 0.31
0.001	38.33 ± 0.49	42.50 ± 1.52	11.67 ± 0.56[c]	8.17 ± 0.31
0.005	42.33 ± 0.80[a]	46.67 ± 1.54[c]	11.50 ± 0.56[c]	7.50 ± 0.56
0.01	48.17 ± 2.09[a]	49.33 ± 1.33b[a]	10.50 ± 0.56[b]	6.67 ± 0.67[c]
0.05	71.33 ± 0.88[a]	74.00 ± 1.00[a]	7.17 ± 2.02[c]	4.67 ± 1.23[b]
0.10	78.33 ± 1.09[a]	80.83 ± 1.66[a]	7.00 ± 1.91[b]	4.33 ± 0.94[b]
0.50	88.33 ± 2.47[a*]	88.33 ± 2.47[a*]	-	-
1.00	97.17 ± 2.96[a*]	97.17 ± 2.96[a*]	-	-

Values are expressed as mean ± SEM of six replicates. a, b, c and d significantly different ($p < 0.001$, $p < 0.01$ and $p < 0.05$ respectively) from control, when t- test was applied.

Straight line regression equation was applied between different concentrations of fenoxycarb and their corresponding growth duration of males and females and longevity of adults to observe the significant correlation:

Growth duration of males	$y = 41.08 + 420.70x$;	$r = 0.95$	$p < 0.01$
Growth duration of females	$y = 44.57 + 408.22x$;	$r = 0.96$	$p < 0.01$
Longevity of male adults	$y = 11.82 - 57.81x$;	$r = -0.89$	$p < 0.05$
Longevity of female adults	$y = 7.84 - 41.39x$;	$r = -0.90$	$p < 0.05$

* Larval tenure since adults did not emerge, so these concentrations have not been accounted in correlation

Effect of Fenoxycarb on the Growth Duration of Fourth Instar Larval Treatment (Table 9 and Figure 15)

It was observed that growth duration for the control moth was 37.83 ± 0.48 days for males and 41.83 ± 1.19 days for females. With the increase of fenoxycarb concentration, in the rearing medium, the growth duration was found to be significantly increased for both the sexes. In case males, the growth duration was recorded to be 38.33 ± 0.49 days at 0.001 ppm concentration of fenoxycarb that was found to be enhanced to 78.33 ± 1.09 days at 0.10 ppm concentration of this JHA. The growth duration of females was observed to be increased 42.50 ± 1.52 days at 0.001 ppm concentration which was further enhanced to 80.83 ± 1.66 days at 0.10 ppm concentration of fenoxycarb. It also deserves mention that at 0.50 and 1.00 ppm concentrations of fenoxycarb larval period was much prolonged which lead to production of giant larvae, supernumerary larvae and their average life span was recorded to be 88.33 ± 2.47 days and 97.17 ± 2.96 days respectively. The larvae

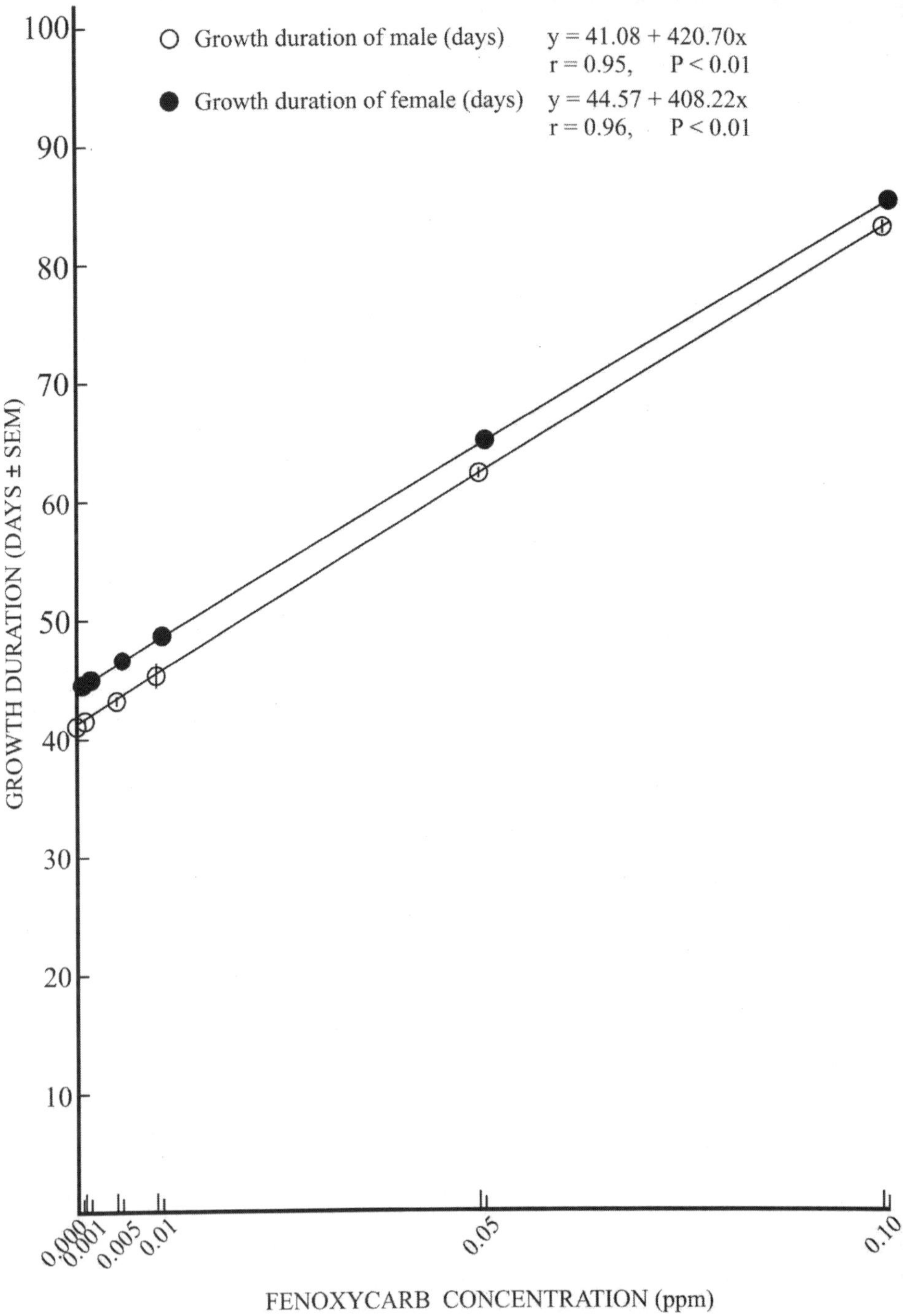

Figure 15: Graphic Representation of Effect of Fenoxycarb on the Growth Duration of the Rice Moth, *Corcyra cephalonica* Exposed as Fourth Instar Larvae (Mean ± SEM).

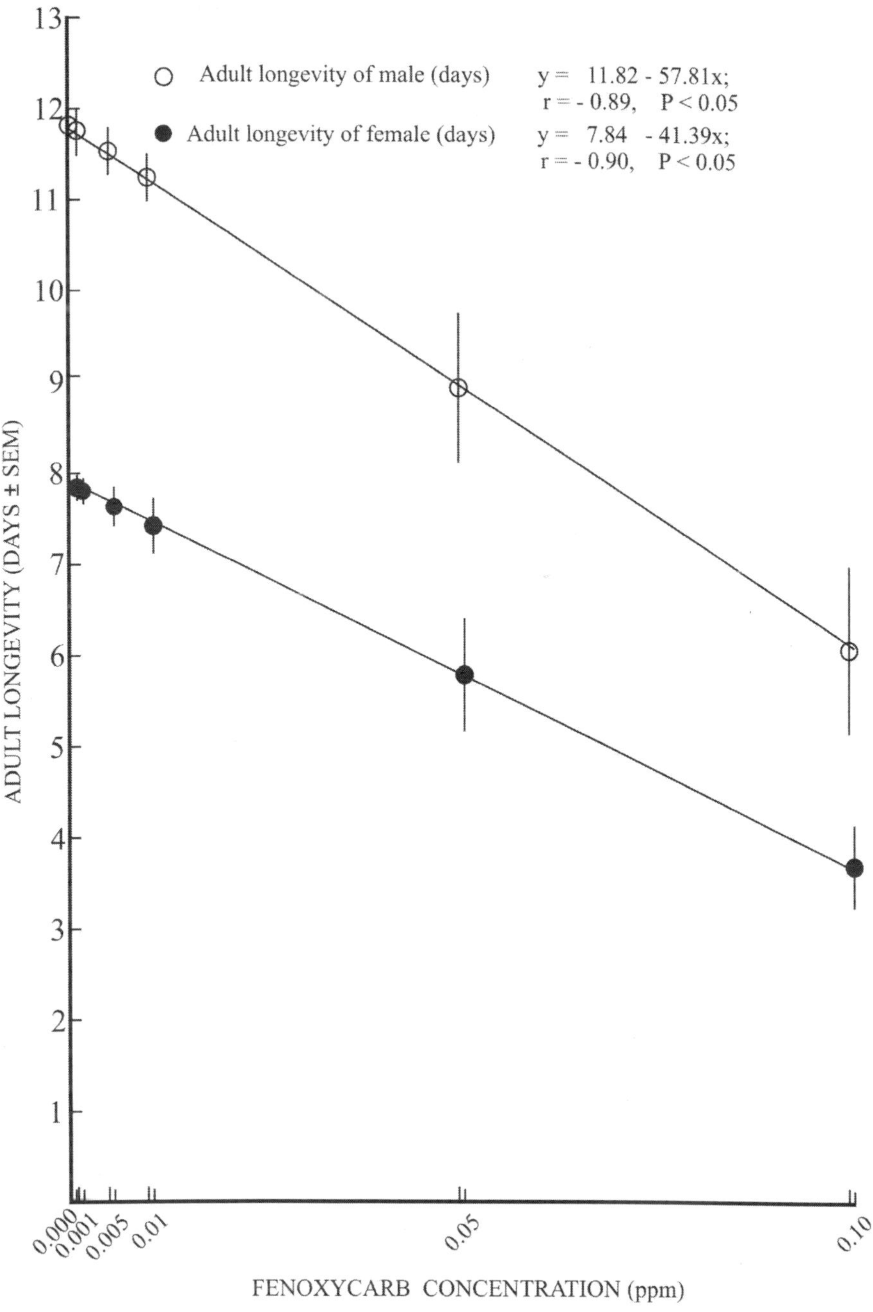

Figure 16: Graphic Representation of Effect of Fenoxycarb on the Adult Longevity of the Rice Moth, *Corcyra cephalonica* Exposed as Fourth Instar Larvae (Mean ± SEM).

at the concentrations of 0.50 and 1.00 ppm fenoxycarb did not emerge as adults so whether these larvae were males or females could not be determined.

Effect of Fenoxycarb on the Logevity of Adults of Fourth Instar Larval Treatment (Table 9 and Figure 16)

The life-span of control moth was recorded to be 13.50 ± 0.43 days for males and 8.83 ± 0.31 days for females. With the increase in the concentrations of this JHA the longevity of males and females individuals were found to be significantly reduced. At 0.001 ppm concentration of fenoxycarb male adult longevity was 11.67 ± 0.56 days while female adult longevity was 8.17 ± 0.31 days. Males and females adults longevity were observed to be 7.00 ± 1.91 and 4.33 ± 0.94 days respectively at 0.10 ppm concentration of fenoxycarb. In addition at 0.05 and 0.10 ppm concentrations of fenoxycarb many of emerged adults were died within 24 h of their emergence.

SECTION I C

Effect of Fenoxycarb on the Egg Laying Capacity (Fecundity) and their Hatchability (Fertility) of Rice Moth, *C. cephalonica*

The total egg output by the females of different crosses and their hatchability at the concentrations of 0.001, 0.005, 0.01and 0.05 was recorded separately and the data, thus obtained, are presented in the Tables 10–13 and Figures 17–24. At 0.10 ppm concentration of fenoxycarb, number of live and capable adults for pairing was not enough, so this concentration was not used for making different crosses. It also deserves mention that treated as well as control females of *C. cephalonica* laid eggs parthenogenetically but such unfertilized eggs did not hatch.

Effect of Fenoxycarb on the Egg Laying Capacity of First Instar Larval Treatment (Table 10 and Figure 17)

It was observed that in control cross (Normal male x Normal female) the total egg output was 495.83 ± 5.16/female. When normal females were crossed with the males emerged from the cultures treated with 0.001, 0.005, 0.01 and 0.05 ppm of fenoxycarb, the total number of eggs laid per female was observed to be 465.67 ± 4.95, 456.17 ± 4.36, 340.33 ± 7.21 and 300.67 ± 6.76 respectively. When normal males were crossed with the females emerged from the cultures, treated with 0.001, 0.005, 0.01 and 0.05 ppm of fenoxycarb, the respective number of eggs per female was recorded as 440.67 ± 4.85, 439.83 ± 4.74, 310.17 ± 7.21 and 270.67 ± 6.84, whereas when the crosses were made between the sexes, both emerged from the cultures treated with 0.001, 0.005, 0.01 and 0.05 ppm of fenoxycarb concentrations, the respective number of eggs per female was recorded as 425.50 ± 4.09, 402.00 ± 4.85, 295.67 ± 6.48 and 220.50 ± 7.24 respectively.

Effect of Fenoxycarb on the Egg Hatchability of First Instar Larval Treatment (Table 10 and Figure 18)

The hatchability of eggs collected from control cross was 92.61 ± 1.10 per cent. The hatchability of eggs, collected from the crosses between normal females and males emerged from the first instar larvae treated with 0.001, 0.005, 0.01 and 0.05 ppm of fenoxycarb concentrations, was found to be 84.90 ± 1.65, 80.22 ± 0.94, 60.17 ± 0.85 and 22.32 ± 2.03 per cent respectively. The hatchability of eggs, collected from the crosses between normal male and the females, emerged from the cultures treated with 0.001, 0.005, 0.01 and 0.05 ppm of fenoxycarb concentrations, was recorded to be 79.17 ± 2.13, 78.72 ± 1.29, 50.42 ± 2.41 and 15.67 ± 2.25 per cent respectively. The hatchability of eggs, collected from the crosses between males and females both emerged from cultures treated with 0.001, 0.005, 0.01 and 0.05 ppm of fenoxycarb concentrations was found to be 73.01± 0.74, 72.13 ± 1.73, 45.13 ± 1.88 and 8.88 ± 1.38 per cent respectively.

Table 10: Effect of Fenoxycarb on the Egg Laying Capacity and their Hatchability of *C. cephalonica* Exposed as First Instar Larvae

Fenoxycarb Concentration (ppm)	*Crossing Sets*	*Fecundity# (Eggs Laid /Female)*	*Fertility# (Eggs Hatched)*	*Per cent Hatchability#*	*Per cent Observed Sterility#*	*Per cent Corrected Sterility#**
Control	N♂ x N♀	495.83 ± 5.16	459.17 ± 6.41	92.61 ± 1.10	7.38 ± 1.10	-
0.001	T♂ x N♀	465.67 ± 4.95[b]	395.00 ± 6.33	84.90 ± 1.65	15.09 ± 1.65	8.32 ± 1.79
	N♂ x T♀	440.67 ± 4.85[a]	348.50 ± 6.97	79.17 ± 2.13[b]	20.83 ± 2.13	14.52 ± 2.30
	T♂ x T♀	425.50 ± 4.09[a]	310.67 ± 4.55	73.01 ± 0.74[a]	26.99 ± 0.74	21.17 ± 0.80
0.005	T♂ x N♀	456.17 ± 4.36[a]	365.83 ± 3.84	80.22 ± 0.94[a]	19.78 ± 0.34	13.39 ± 1.01
	N♂ x T♀	439.83 ± 4.74[a]	346.17 ± 5.86	78.72 ± 1.29[a]	21.28 ± 1.29	15.00 ± 1.39
	T♂ x T♀	402.00 ± 4.85[a]	289.67 ± 4.79	72.13 ± 1.73[a]	27.87 ± 1.73	22.12 ± 1.87
0.01	T♂ x N♀	340.33 ± 7.21[a]	204.50 ± 2.81	60.17 ± 0.85[a]	39.84 ± 0.85	35.05 ± 0.92
	N♂ x T♀	310.17 ± 7.21[a]	156.00 ± 6.45	50.42 ± 2.41[a]	49.58 ± 2.41	45.56 ± 2.60
	T♂ x T♀	295.67 ± 6.48[a]	133.17 ± 4.88	45.13 ± 1.88[a]	54.87 ± 1.88	51.27 ± 2.03
0.05	T♂ x N♀	300.67 ± 6.76[a]	67.50 ± 7.21	22.32 ± 2.03[a]	77.68 ± 2.03	75.90 ± 2.18
	N♂ x T♀	270.67 ± 6.84[a]	42.67 ± 6.50	15.67 ± 2.25[a]	84.33 ± 2.25	83.08 ± 2.43
	T♂ x T♀	220.50 ± 7.24[a]	19.17 ± 2.45	8.88 ± 1.38[a]	91.12 ± 1.38	90.42 ± 1.50

Values are expressed as mean ± SEM of six replicates. a and b Significantly different $p < 0.001$ and $p < 0.01$ respectively compared with control when t- test was applied. * Calculated by Chamberlain's formula (1962).

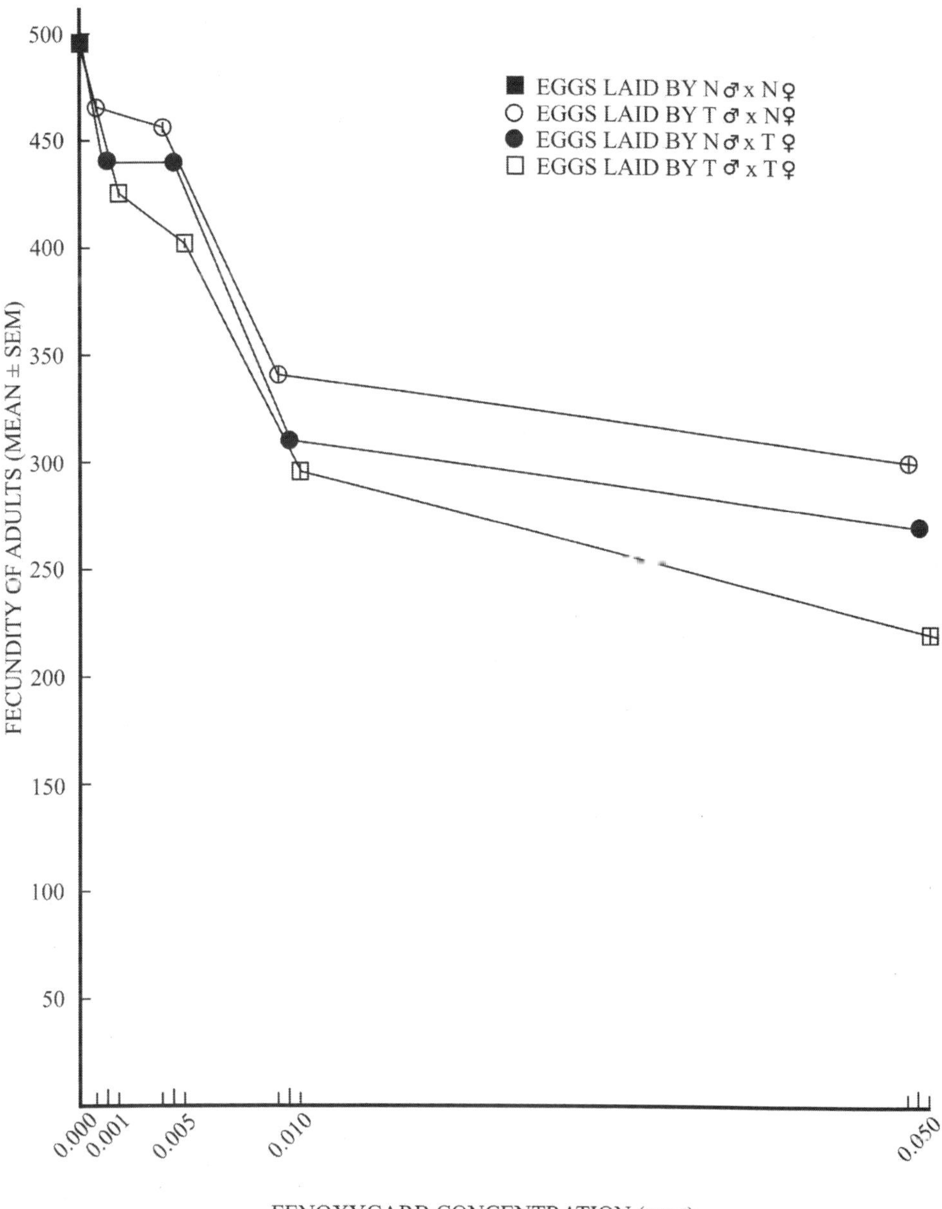

Figure 17: Graphic Representation of Effect of Fenoxycarb on the Fecundity of Adults of the Rice Moth, *Corcyra cephalonica* Exposed as First Instar Larvae (Mean ± SEM).

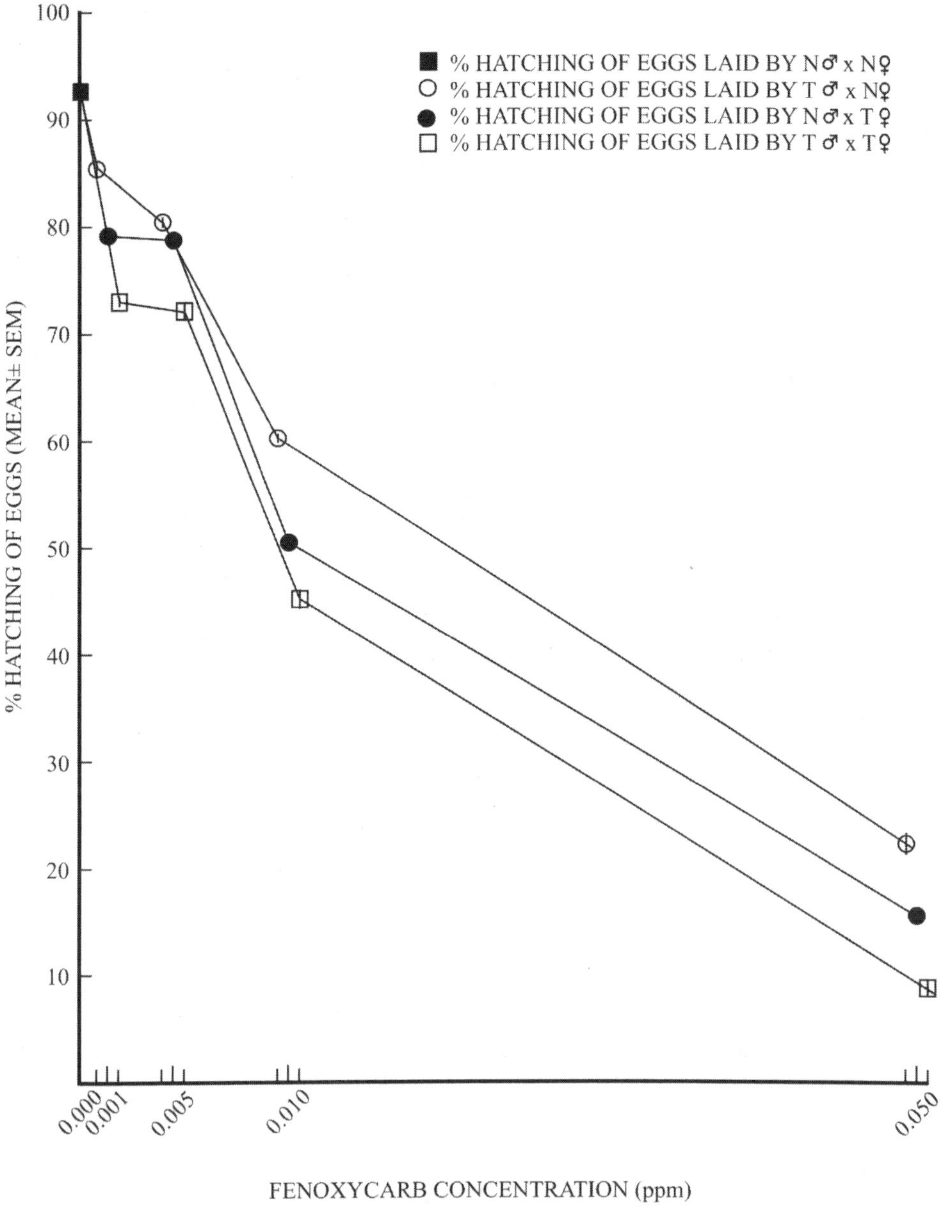

Figure 18: Graphic Representation of Effect of Fenoxycarb on the Per cent Hatching of Eggs Laid by the Rice Moth, *Corcyra cephalonica* Exposed as First Instar Larvae (Mean ± SEM).

Effect of Fenoxycarb on the Egg Laying Capacity of Second Instar Larval Treatment (Table 11 and Figure 19)

It was observed that in control cross (Normal male x Normal female) the total egg output was 492.17 ± 3.60/female. When normal females were crossed with the males emerged from the cultures treated with 0.001, 0.005, 0.01 and 0.05 ppm of fenoxycarb concentrations, the respective number of eggs laid per female was found to be 473.83 ± 6.79, 461.83 ± 6.92, 384.50 ± 6.72 and 369.50 ± 7.30. When normal males were crossed with the females emerged from the cultures, treated with the same concentrations *i.e.* 0.001, 0.005, 0.01 and 0.05 ppm of fenoxycarb, the number of eggs laid per female was recorded as 461.33 ± 6.82, 445.50 ± 5.17, 360.50 ± 6.85 and 330.50 ± 4.81 respectively. When males and females, both emerged from the cultures treated with the concentrations 0.001, 0.005, 0.01 and 0.05 ppm of fenoxycarb were crossed, the respective number of eggs laid per female was observed to be 435.33 ± 7.04, 410.67 ± 6.19, 340.33 ± 5.91 and 286.17 ± 6.15.

Effect of Fenoxycarb on the Egg Hatchability of Sedcond Instar Larval Treatment (Table 11 and Figure 20)

The hatchability of eggs collected from control cross was found to be 94.23 ± 0.54 per cent.The hatchability of eggs, collected from the crosses between normal females and males emerged from the cultures treated with 0.001, 0.005, 0.01 and 0.05 ppm of fenoxycarb concentrations, was found to be 85.02 ± 1.08, 79.32 ± 1.42, 61.77 ± 0.63 and 24.74 ± 1.15 per cent respectively. The hatchability of eggs, collected from the crosses between normal male and the females, emerged from the cultures treated with 0.001, 0.005, 0.01 and 0.05 ppm of fenoxycarb concentrations, was recorded to be 81.13 ± 1.26, 75.24 ± 1.22, 58.17 ± 1.67 and 16.00 ± 1.00 per cent respectively. The hatchability of eggs, collected from the crosses between males and females both emerged from cultures treated with 0.001, 0.005, 0.01 and 0.05 ppm of fenoxycarb concentrations was observed to be 77.36 ± 1.05, 71.56 ± 1.29, 50.39 ± 1.15 and 10.63 ± 0.80 per cent respectively.

Effect of Fenoxycarb on the Egg Laying Capacity of Third Instar Larval Treatment (Table 12 and Figure 21)

In third instar larval treatment the egg laying capacity of control cross (Normal male X Normal female) was found to be 494.50 ± 3.89/female. When normal females, were crossed with the males emerged from third instar larvae, treated with 0.001, 0.005, 0.01 and 0.05 ppm concentration of fenoxycarb, the respective number of eggs laid per female was recorded to be 485.17 ± 5.25, 458.00 ± 5.52, 432.50 ± 6.84 and 382.33 ± 4.74. When normal males were crossed with the females emerged from the cultures, treated with 0.001, 0.005, 0.01 and 0.05 ppm concentrations of fenoxycarb, the total number of eggs laid per female was found to be 472.50 ± 4.99, 445.67 ± 5.49, 421.33 ± 7.26 and 343.33 ± 6.02 respectively. When both the sexes used in the cross were emerged from the cultures treated with 0.001, 0.005, 0.01 and 0.05 ppm concentrations of fenoxycarb, the respective number of eggs laid per female was recorded to be 442.67 ± 5.12, 418.00 ± 6.28, 380.67 ± 5.95 and 310.17 ± 7.60.

Table 11: Effect of Fenoxycarb on the Egg Laying Capacity and their Hatchability of *C. cephalonica* Exposed as Second Instar Larvae

Fenoxycarb Concentration (ppm)	*Crossing Sets*	*Fecundity# (Eggs Laid /Female)*	*Fertility# (Eggs Hatched)*	*Per cent Hatchability#*	*Per cent Observed Sterility#*	*Per cent Corrected Sterility#**
Control	N♂ x N♀	492.17 ± 3.60	463.83 ± 5.15	94.23 ± 0.54	5.77 ± 0.54	-
0.001	T♂ x N♀	473.83 ± 6.79	402.83 ± 7.98	85.02 ± 1.08[a]	14.98 ± 1.18	9.78 ± 1.25
	N♂ x T♀	461.33 ± 6.82[b]	374.00 ± 4.26	81.13 ± 1.26[a]	18.87 ± 1.26	13.90 ± 1.34
	T♂ x T♀	435.33 ± 7.04[a]	336.67 ± 5.72	77.36 ± 1.05[a]	22.64 ± 1.05	17.90 ± 1.11
0.005	T♂ x N♀	461.83 ± 6.92[b]	366.17 ± 7.14	79.32 ± 1.42[a]	20.68 ± 1.42	15.83 ± 1.39
	N♂ x T♀	445.50 ± 5.17[a]	335.33 ± 8.01	75.24 ± 1.22[a]	24.76 ± 1.22	20.16 ± 1.30
	T♂ x T♀	410.67 ± 6.19[a]	293.67 ± 4.51	71.56 ± 1.29[a]	28.44 ± 1.29	24.06 ± 1.37
0.01	T♂ x N♀	384.50 ± 6.72[a]	237.33 ± 2.73	61.77 ± 0.63[a]	38.23 ± 0.63	34.45 ± 0.67
	N♂ x T♀	360.50 ± 6.85[a]	209.50 ± 5.78	58.17 ± 1.67[a]	41.83 ± 1.67	38.27 ± 1.77
	T♂ x T♀	340.33 ± 5.91 [a]	171.33 ± 3.39	50.39 ± 1.15[a]	49.61 ± 1.15	46.52 ± 1.22
0.05	T♂ x N♀	369.50 ± 7.30[a]	91.50 ± 4.86	24.74 ± 1.15[a]	75.26 ± 1.15	73.74 ± 1.22
	N♂ x T♀	330.50 ± 4.81[a]	53.00 ± 3.68	16.00 ± 1.00[a]	84.00 ± 1.00	83.02 ± 1.06
	T♂ x T♀	286.17 ± 6.15[a]	30.33 ± 2.11	10.63 ± 0.80[a]	89.37 ± 0.80	88.72 ± 0.85

Values are expressed as mean ± SEM of six replicates. a and b Significantly different $p < 0.001$ and $p < 0.01$ respectively compared with control when t- test was applied. * Calculated by Chamberlain's formula (1962).

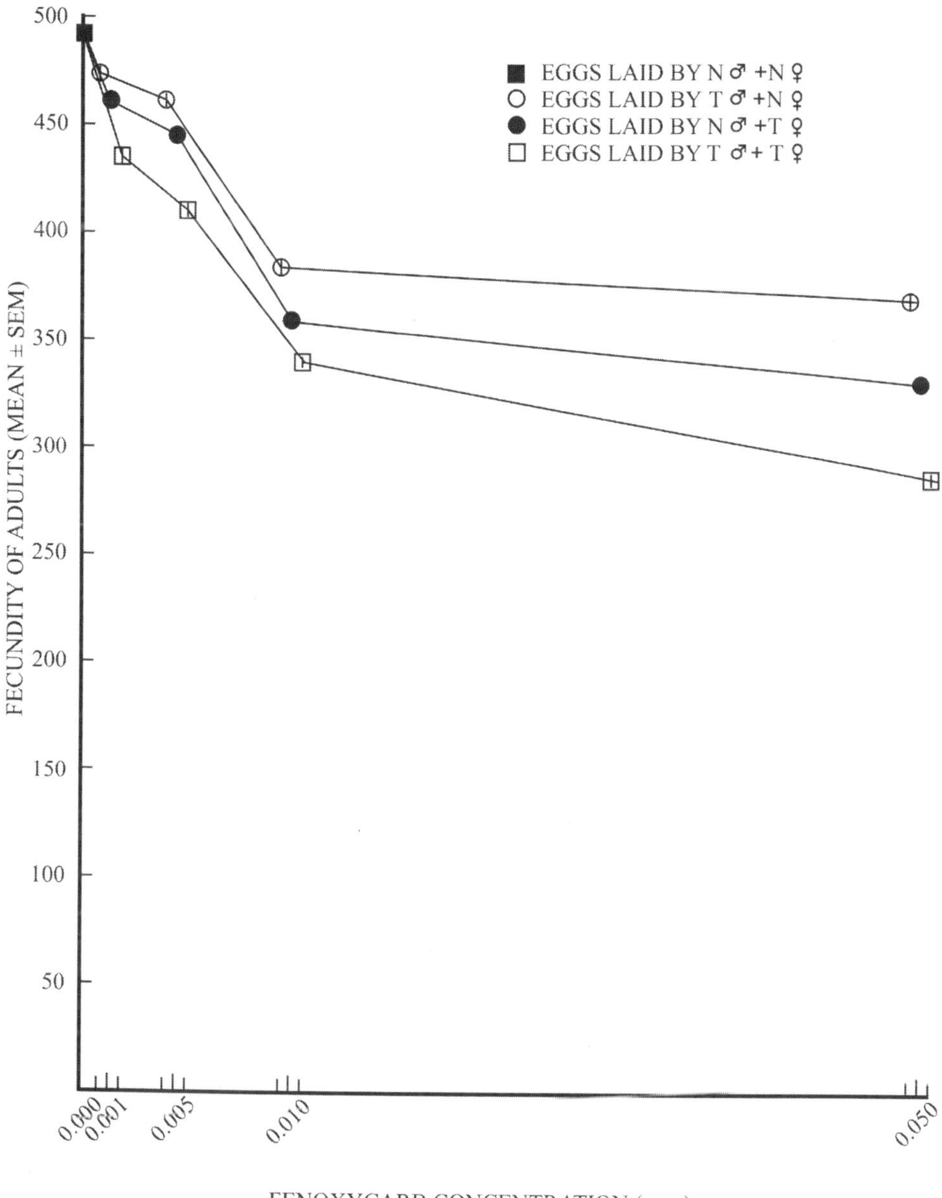

Figure 19: Graphic Representation of Effect of Fenoxycarb on the Fecundity of Adults of the Rice Moth, *Corcyra cephalonica* Exposed as Second Instar Larvae (Mean ± SEM).

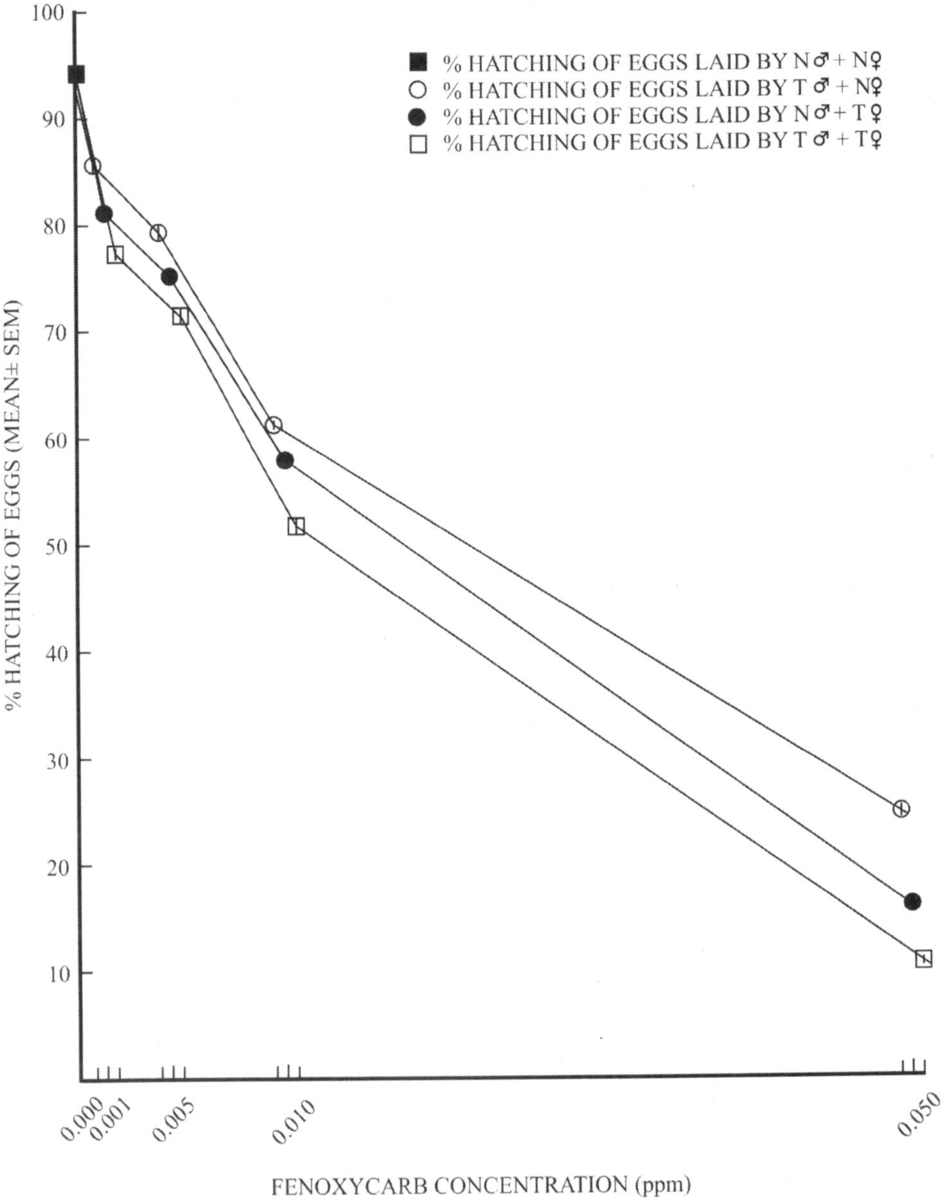

Figure 20: Graphic Representation of Effect of Fenoxycarb on the Per cent Hatching of Eggs Laid by the Rice Moth, *Corcyra cephalonica* Exposed as Second Instar Larvae (Mean ± SEM).

Table 12: Effect of Fenoxycarb on the Egg Laying Capacity and their Hatchability of *C. cephalonica* Exposed as Third Instar Larvae

Fenoxycarb Concentration (ppm)	*Crossing Sets*	*Fecundity# (Eggs Laid /Female)*	*Fertility# (Eggs Hatched)*	*Per cent Hatchability#*	*Per cent Observed Sterility#*	*Per cent Corrected Sterility#**
Control	N♂ x N♀	494.50 ± 3.89	469.00 ± 7.375	94.82 ± 0.90	5.18 ± 0.90	
0.001	T♂ x N♀	485.17 ± 5.25	418.17 ± 5.98	86.18 ± 0.48[a]	13.82 ± 0.48	9.11 ± 0.50
	N♂ x T♀	472.50 ± 4.99[b]	397.00 ± 1.23	84.07 ± 0.91[a]	15.93 ± 0.91	11.34 ± 0.96
	T♂ x T♀	442.67 ± 5.12[a]	363.67 ± 7.72	82.15 ± 1.35[a]	17.85 ± 1.35	13.37 ± 1.42
0.005	T♂ x N♀	458.00 ± 5.52[a]	374.67 ± 3.25	81.85 ± 0.96[a]	18.15 ± 0.96	13.68 ± 1.00
	N♂ x T♀	445.67 ± 5.49[a]	357.00 ± 7.38	80.11 ± 1.43[a]	19.89 ± 1.43	15.51 ± 1.51
	T♂ x T♀	418.00 ± 6.28[a]	331.17 ± 7.22	79.25 ± 1.57[a]	20.75 ± 1.57	16.42 ± 1.66
0.01	T♂ x N♀	432.50 ± 6.84[a]	281.67 ± 2.40	65.17 ± 0.73[a]	34.83 ± 0.73	31.27 ± 0.77
	N♂ x T♀	421.33 ± 7.26[a]	266.00 ± 6.23	63.11 ± 0.70[a]	36.89 ± 0.70	33.44 ± 0.74
	T♂ x T♀	380.67 ± 5.95[a]	228.33 ± 4.75	60.00 ± 1.14[a]	39.99 ± 1.14	36.71 ± 1.20
0.05	T♂ x N♀	382.33 ± 4.74[a]	96.67 ± 2.85	25.33 ± 1.03[a]	74.66 ± 1.03	73.28 ± 1.08
	N♂ x T♀	343.33 ± 6.02[a]	63.17 ± 7.09	18.41 ± 2.06[a]	81.41 ± 2.06	80.60 ± 2.17
	T♂ x T♀	310.17 ± 7.60[a]	35.17 ± 2.44	11.28 ± 0.56[a]	88.72 ± 0.56	88.10 ± 0.59

Values are expressed as mean ± SEM of six replicates. a and b Significantly different $p < 0.001$ and $p < 0.01$ respectively compared with control when t- test was applied. * Calculated by Chamberlain's formula (1962).

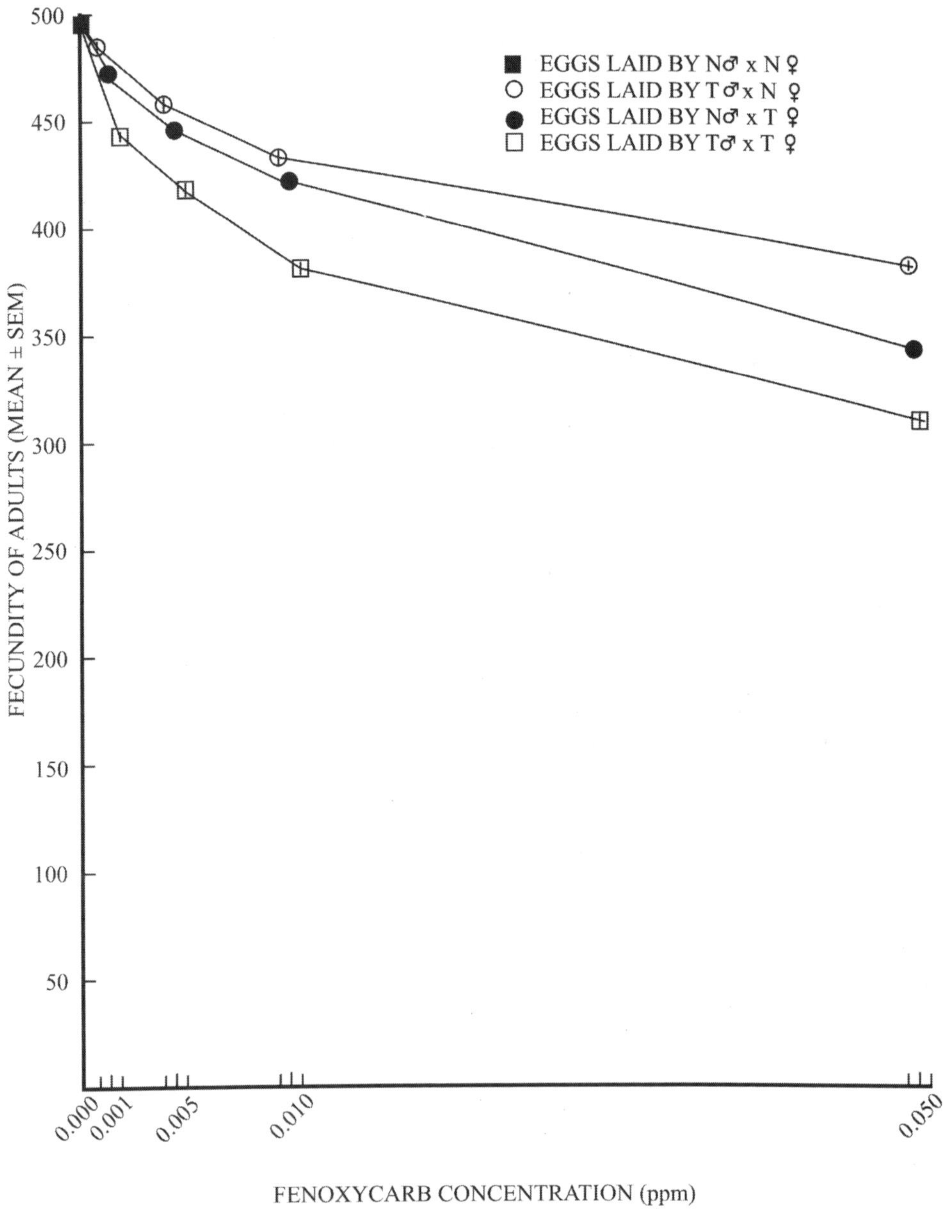

Figure 21: Graphic Representation of Effect of Fenoxycarb on the Fecundity of Adults of the Rice Moth, *Corcyra cephalonica* Exposed as Third Instar Larvae (Mean ± SEM).

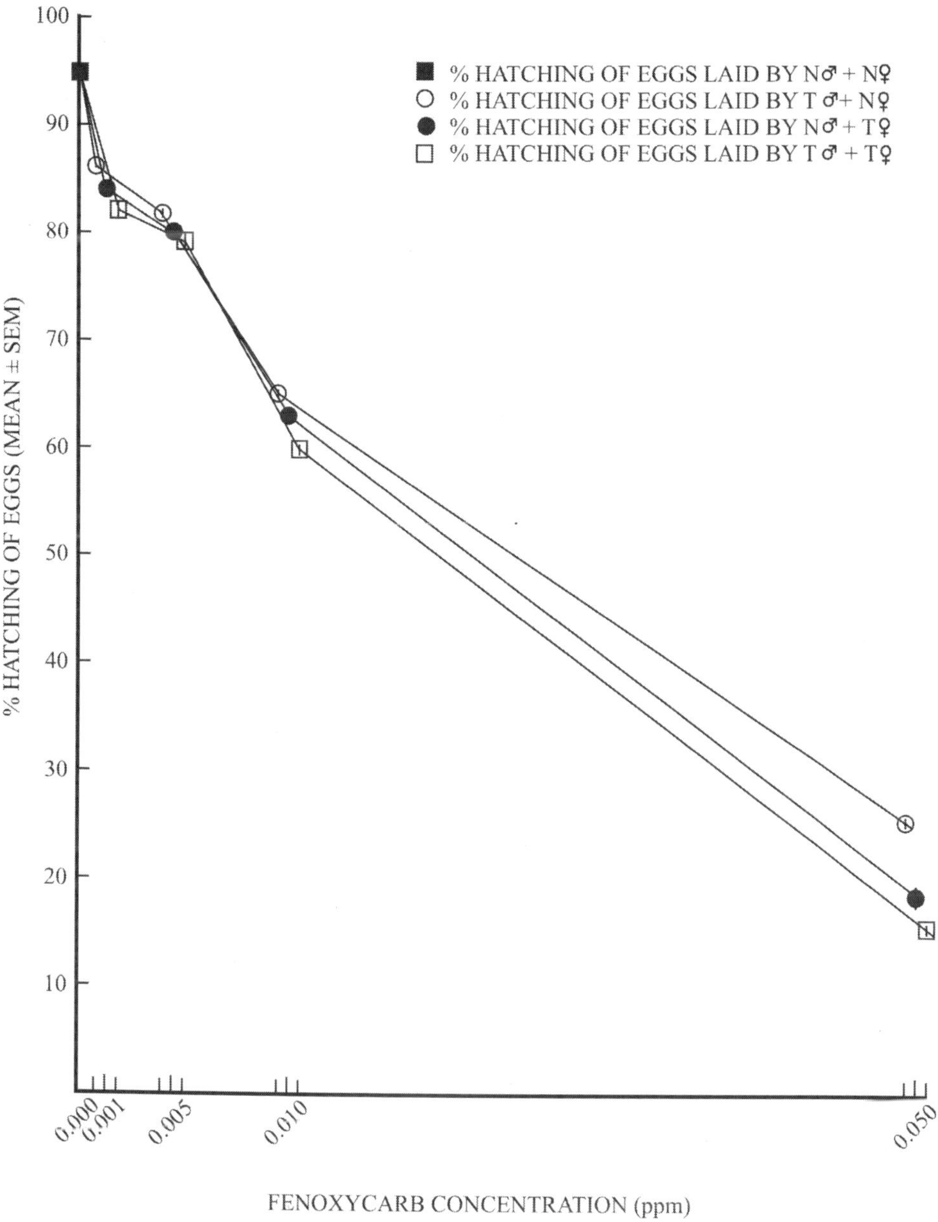

Figure 22: Graphic Representation of Effect of Fenoxycarb on the Per cent Hatching of Eggs Laid by the Rice Moth, *Corcyra cephalonica* Exposed as Third Instar Larvae (Mean ± SEM).

Effect of Fenoxycarb on the Egg Hatchability of Third Instar Larval Treatment (Table 12 and Figure 22)

The hatchability of eggs collected from the crosses between normal male and normal female was 94.82 ± 0.92 per cent. The hatchability of eggs, collected from the crosses between normal females and males emerged from the cultures treated with 0.001, 0.005, 0.01 and 0.05 ppm of fenoxycarb concentrations, was found to be 86.18 ± 0.48, 81.85 ± 0.96, 65.17 ± 0.73 and 25.33 ± 1.03 per cent respectively. The hatchability of eggs, collected from the crosses between normal male and the females, emerged from the cultures treated with 0.001, 0.005, 0.01 and 0.05 ppm of fenoxycarb concentrations, was observed to be 84.07 ± 0.91, 80.11 ± 1.43, 63.11 ± 0.70 and 18.41 ± 2.06 per cent respectively. The hatchability of eggs, collected from the crosses, where both sexes were emerged from the cultures treated with 0.001, 0.005, 0.01 and 0.05 ppm of fenoxycarb concentrations, was recorded to be 82.15 ± 1.35, 79.25 ± 1.57, 60.00 ± 1.14 and 11.28 ± 0.56 per cent respectively.

Effect of Fenoxycarb on the Egg Laying Capacity of Fourth Instar Larval Treatment (Table 13 and Figure 23)

It was observed that in control cross (Normal male x Normal female) the total egg output was 493.00 ± 4.44/female. When normal females were crossed with the males emerged from the fourth instar larvae treated with 0.001, 0.005, 0.01 and 0.05 ppm of fenoxycarb concentrations, the number of eggs laid per female was found to be 485.17 ± 6.22, 460.50 ± 5.07, 435.17 ± 6.67 and 392.17 ± 4.25 respectively. when normal males were crossed with the females emerged from the cultures, treated with 0.001, 0.005, 0.01 and 0.05 ppm concentrations of fenoxycarb, the respective number of eggs laid per female was recorded to be 475.83 ± 5.64, 457.17 ± 6.86, 421.67 ± 7.61 and 351.50 ± 7.25. When both the sexes used in the cross were emerged from the cultures treated with 0.001, 0.005, 0.01 and 0.05 ppm concentrations of fenoxycarb, the respective number of eggs laid per female was observed to be 445.67 ± 6.18, 438.17 ± 6.39, 397.00 ± 7.26 and 320.83 ± 7.25.

Effect of Fenoxycarb on the Egg Hatchability of Fourth Instar Larval Treatment (Table 13 and Figure 24)

The hatchability of eggs collected from control cross was found to be 93.59 ± 0.77 per cent. In the eggs, collected from the crosses between normal females and the males emerged from fourth instar larvae, treated with 0.001, 0.005, 0.01 and 0.05 ppm of fenoxycarb concentrations, was recorded to be 89.92 ± 0.94, 84.04 ± 0.96, 75.57 ± 1.87 and 50.38 ± 1.41 per cent respectively. The hatchability of eggs, collected from the crosses between normal male and the females, emerged from the cultures treated with 0.001, 0.005, 0.01 and 0.05 ppm of fenoxycarb concentrations, was observed to be 88.62 ± 1.36, 82.38 ± 1.23, 72.39 ± 0.64 and 44.32 ± 1.25 per cent respectively. The hatchability of eggs, collected from the crosses between males and females both emerged from cultures treated with 0.001, 0.005, 0.01 and 0.05 ppm of fenoxycarb concentration was found to be 85.46 ± 0.78, 80.10 ± 2.42, 66.76 ± 2.02 and 22.53 ± 1.73 per cent respectively.

Table 13: Effect of Fenoxycarb on the Egg Laying Capacity and their Hatchability of *C. cephalonica* Exposed as Fourth Instar Larvae

Fenoxycarb Concentration (ppm)	*Crossing Sets*	*Fecundity# (Eggs Laid /Female)*	*Fertility# (Eggs Hatched)*	*Per cent Hatchability#*	*Per cent Observed Sterility#*	*Per cent Corrected Sterility#**
Control	N♂ x N♀	493.00 ± 4.44	461.33 ± 3.87	93.59 ± 0.77	6.41 ± 0.77	-
0.001	T♂ x N♀	485.17 ± 6.22	436.00 ± 3.26	89.92 ± 0.94[c]	10.08 ± 0.94	3.93 ± 1.01
	N♂ x T♀	475.83 ± 5.64[c]	421.33 ± 3.52	88.62 ± 1.36[b]	11.38 ± 1.36	5.32 ± 1.45
	T♂ x T♀	445.67 ± 6.18[a]	380.67 ± 3.52	85.46 ± 0.78[a]	14.54 ± 0.78	8.70 ± 0.83
0.005	T♂ x N♀	460.50 ± 5.07[a]	387.00 ± 5.87	84.04 ± 0.96[a]	15.96 ± 0.96	10.20 ± 1.02
	N♂ x T♀	457.17 ± 6.86[b]	376.33 ± 4.25	82.38 ± 1.23[a]	17.62 ± 1.23	11.98 ± 1.32
	T♂ x T♀	438.17 ± 6.39[a]	350.33 ± 6.70	80.10 ± 2.42[a]	19.90 ± 2.42	14.41 ± 2.58
0.01	T♂ x N♀	435.17 ± 6.67[a]	328.33 ± 4.63	75.57 ± 1.87[a]	24.43 ± 1.87	19.26 ± 2.00
	N♂ x T♀	421.67 ± 7.61[a]	305.17 ± 5.24	72.39 ± 0.64[a]	27.61 ± 0.64	22.66 ± 0.68
	T♂ x T♀	397.00 ± 7.26[a]	264.67 ± 7.09	66.76 ± 2.02[a]	33.24 ± 2.02	28.67 ± 2.16
0.05	T♂ x N♀	392.17 ± 4.25[a]	197.33 ± 4.11	50.38 ± 1.41[a]	49.62 ± 1.41	46.17 ± 1.50
	N♂ x T♀	351.50 ± 7.25[a]	156.00 ± 6.48	44.32 ± 1.25[a]	55.68 ± 1.25	52.64 ± 1.34
	T♂ x T♀	320.83 ± 7.25[a]	72.00 ± 5.17	22.53 ± 1.73[a]	77.47 ± 1.73	75.93 ± 1.84

Values are expressed as mean ± SEM of six replicates. a, b and c Significantly different $p < 0.001$, $p < 0.01$ and $p < 0.05$ respectively compared with control when t- test was applied. * Calculated by Chamberlain's formula (1962).

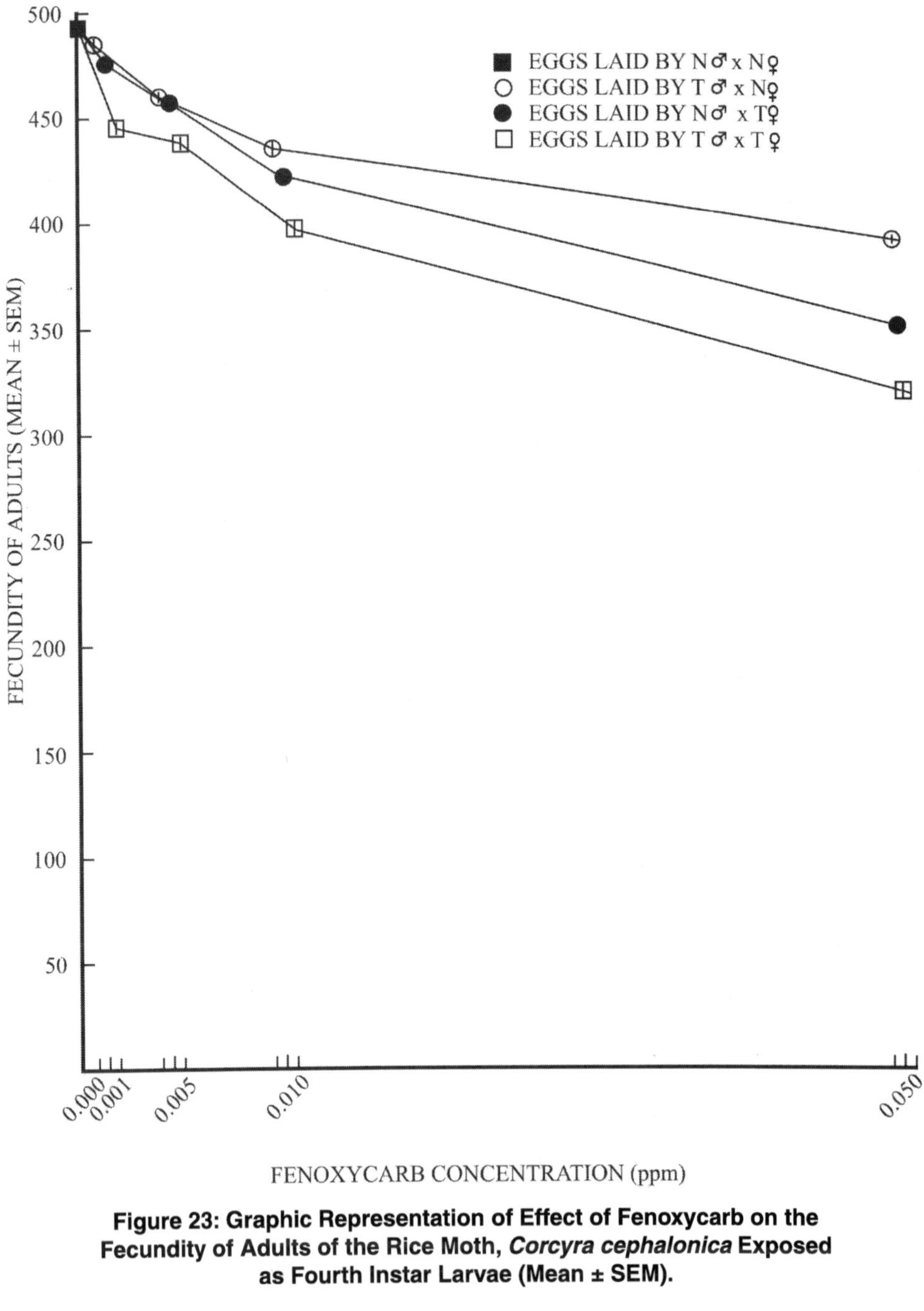

Figure 23: Graphic Representation of Effect of Fenoxycarb on the Fecundity of Adults of the Rice Moth, *Corcyra cephalonica* Exposed as Fourth Instar Larvae (Mean ± SEM).

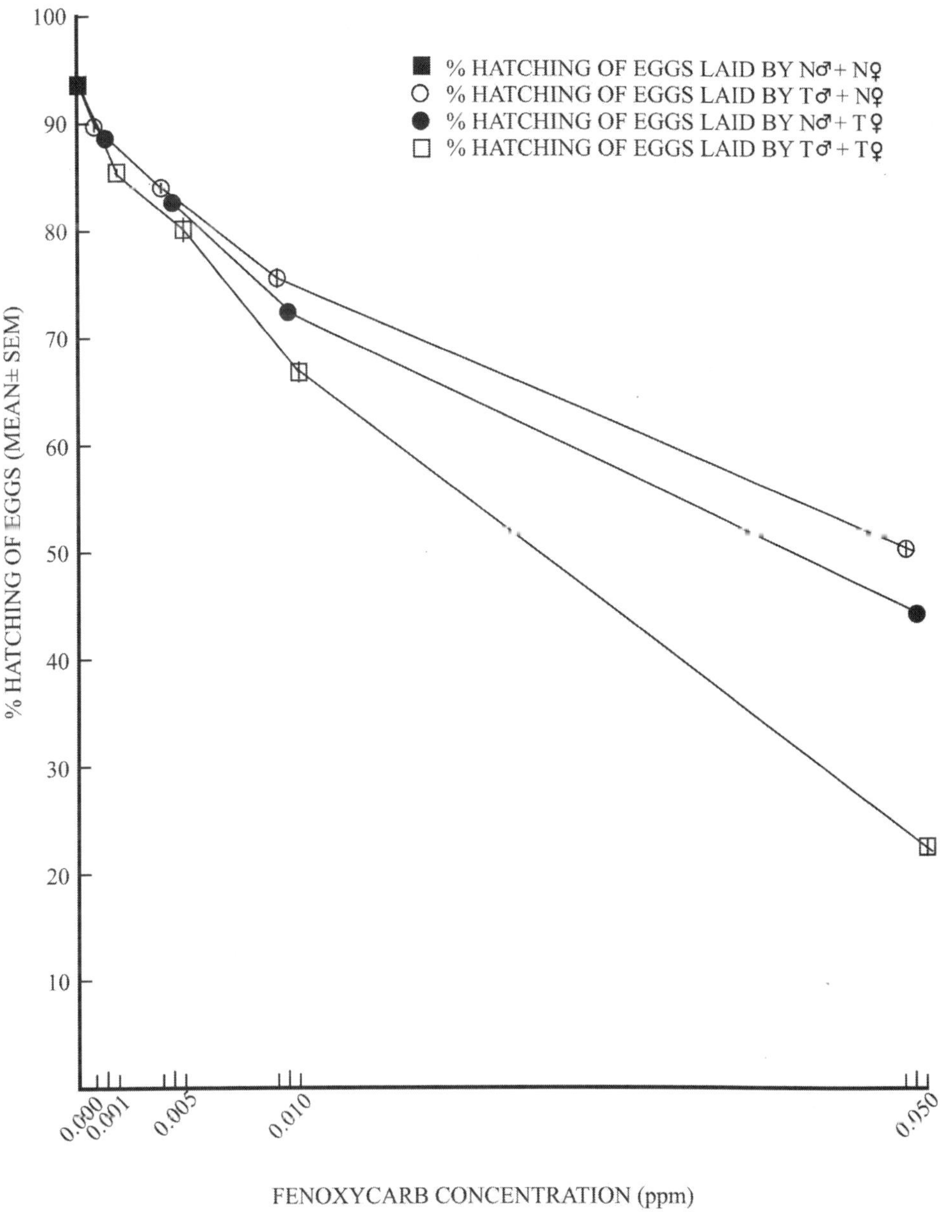

Figure 24: Graphic Representation of Effect of Fenoxycarb on the Per cent Hatching of Eggs Laid by the Rice Moth, *Corcyra cephalonica* Exposed as Fourth Instar Larvae (Mean ± SEM).

It deserves mention that the data given in table 10 to 13 indicated that the egg laying capacity (fecundity) and hatchability (fertility) is low when both the sexes of pair were treated. However, the rate of fecundity and fertility was decreased when only female of the pair was treated, in comparison to those crosses where only male of the pair was treated. It was observed that at least one treated adult of the pair either male or female will cause the pair had low fecundity and fertility. In addition, it is also noticeable that in the present investigation fecundity as well as fertility of the eggs was lowest in the first instar larval treatment which gradually increased in second, third and fourth instar larval treatments. Lowest fecundity as well as fertility, in the first instar treated larvae, was ofcourse due to their lenthy tenure of exposure to fenoxycarb. In the successive instars (second, third and fourth instars) the tenure of fenoxycarb exposure decreased and hence their fecundity and fertility were less suppressed in comparison to first instar exposed larvae.

SECTION II – BIOCHEMICAL STUDIES

Effect of Fenoxycarb on the Biochemical Constituent of Gonadial Tissues (Testes and Ovaries of the Adults) and Eggs (Laid within 24 hour of Mating) of Rice Moth, *Corcyra cephalonica* Exposed as First Instar Larvae

In this section the effect of fenoxycarb on the levels of total protein, total free amino acids, DNA, RNA, glycogen and activities of acid and alkaline phosphatases in testes and ovaries of adults and eggs laid within 24 h of mating by the females of all the four crosses (*i.e.* Normal male x Normal female, Treated male x Normal female, Normal male x Treated female and Treated male x Treated female) of *C. cephalonica* exposed as first instar larvae have been incorporated.

Total Protein Level in the Testes (Table 14, Figure 25)

Sub lethal concentrations of fenoxycarb caused a significantly dose-dependent ($p < 0.001$) reduction in the level of total protein in the testes of rice moth, *C. cephalonica*. In case of control insects, the total protein content in the testes was 34.62 ± 0.38 μg/mg. The maximum decrease in total protein level in testes (49 per cent of the control value) was observed in moths emerged from treated culture of 0.010 ppm concentration of fenoxycarb. Total protein levels in testes were reduced to 80 per cent (27.70 ± 0.47 μg/mg), 67 per cent (23.20 ± 0.98 μg/mg) and 49 per cent (16.96 ± 0.66 μg/mg) of the control value following exposure of 0.001, 0.005 and 0.010 ppm concentrations of fenoxycarb respectively.

Table 14: Changes in the Total Protein and Total Free Amino Acid Levels in the Testes and Ovaries of the Rice Moth, *C. cephalonica* Treated with Fenoxycarb

Fenoxycarb Concentration (ppm)	*Total Protein[#] (μg/mg, wet wt.)*		*Total Free Amino Acid[#] (μg/mg, wet wt.)*	
	Testes	*Ovaries*	*Testes*	*Ovaries*
Control	34.62 ± 0.38 (100)	46.12 ± 0.45 (100)	16.07 ± 0.41 (100)	20.02 ± 0.36 (100)
0.001	27.70 ± 0.47[a] (80)	37.36 ± 0.98[a] (81)	18.32 ± 0.47[b] (114)	21.82 ± 0.86 (109)
0.005	23.20 ± 0.98[ab'] (67)	22.60 ± 0.69[aa'] (49)	23.30 ± 0.95[aa'] (145)	25.03 ± 1.09[bc'] (125)
0.010	16.96 ± 0.66[aa'] (49)	16.60 ± 0.46[aa'] (36)	27.16 ± 1.09[ac'] (169)	35.44 ± 1.31[aa'] (177)

Values are expressed as the mean ± SEM of six replicates; the values in the parentheses indicate the percentage change with control values taken as 100 per cent. a and b Significantly different $p < 0.001$ and $p < 0.01$ respectively compared with controls; a', b' and c' Significantly different $p < 0.001$, $p < 0.01$ and $p < 0.05$ respectively from the corresponding treated group when t-test was applied. Analysis of variance showed that the response to the JHA was dose-dependent $p < 0.001$.

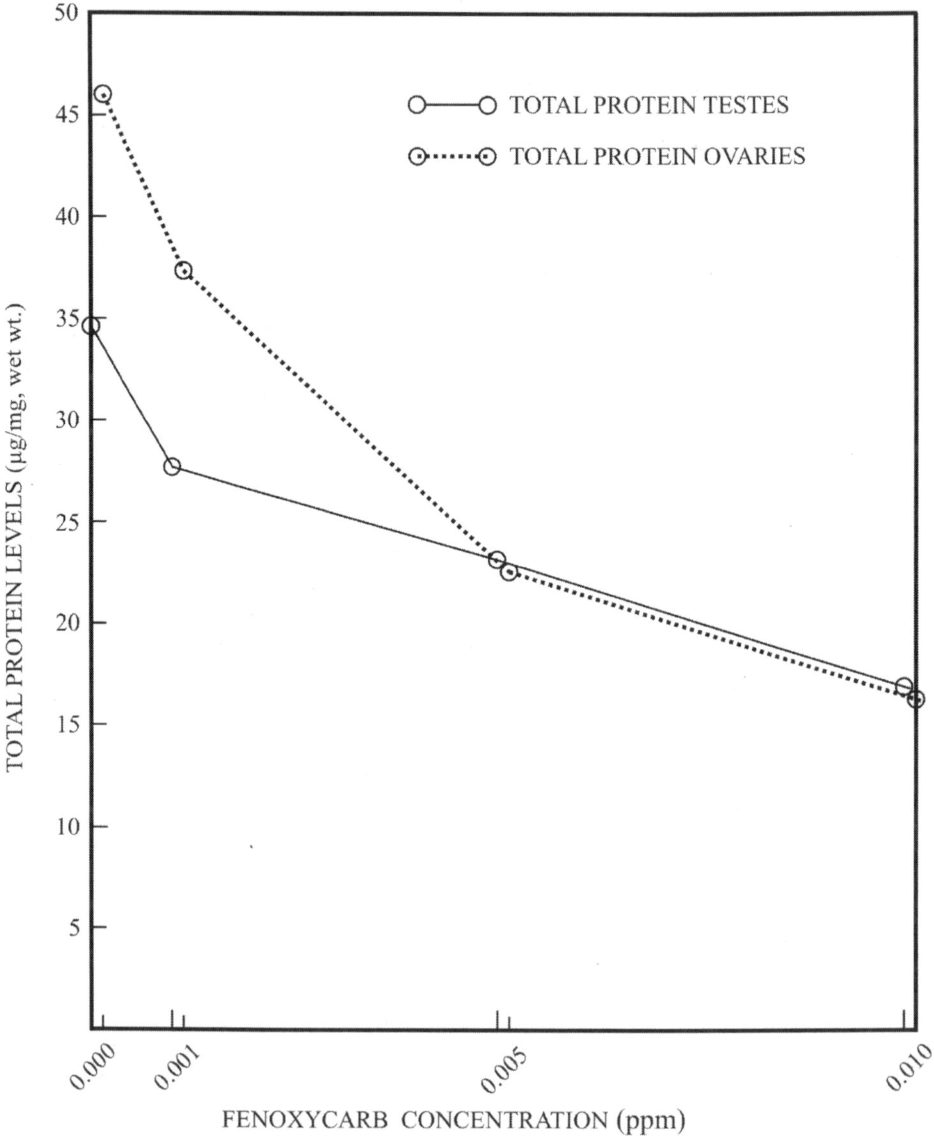

Figure 25: Graphic Representation of Effect of Fenoxycarb on the Total Protein Levels in Tests and Ovaries of the Adult Rice Moth, *Corcyra cephalonica*.

Total Protein Level in the Ovaries (Table 14, Figure 25)

Fenoxycarb at its sub lethal concentrations caused a significantly dose-dependent ($p < 0.001$) reduction in the level of total protein in the ovaries of rice moth, *C. cephalonica*. The total protein level in the ovaries of control insect, was 46.12 ± 0.45 µg/mg. The maximum decrease in total protein level in ovaries (36 per cent of the control value) was recorded in adults emerged from treated culture with 0.010 ppm concentration of fenoxycarb. Total protein levels in ovaries, were reduced to 81 per cent (37.36 ± 0.98 µg/mg), 49 per cent (22.60 ± 0.69 µg/mg) and 36 per cent (16.60 ± 0.46 µg/mg) of the control value following treatment with 0.001, 0.005 and 0.010 ppm concentrations of fenoxycarb respectively.

Total Protein Level in the Eggs (Table 15, Figure 27)

Exposure of sub lethal concentrations of fenoxycarb caused a significantly dose-dependent ($p < 0.001$) reduction in total protein level in the eggs, collected from different crosses of *C. cephalonica*. The total protein content in the eggs, obtained from control cross (normal males and normal females), was recorded to be 1.78 ± 0.14 µg/egg. The maximum decrease in total protein level in the eggs, collected from the crosses of treated males and normal females, normal males and treated females, and treated males and treated females, was recorded to be 57, 52 and 37 per cent of the control value respectively following treatment with 0.010 ppm concentration of fenoxycarb.

Table 15: Effect of Fenoxycarb on the Total Protein Level (µg/egg) in the Eggs of *Corcyra cephalonica*[#]

Fenoxycarb Concentration (ppm)	*Treated Male x Normal Female*	*Normal Male x Treated Female*	*Treated Male x Treated Female*
0.00 (control)	1.78 ± 0.14 (100)	–	–
0.001	1.43 ± 0.08 (80)	1.37 ± 0.06[c] (77)	1.26 ± 0.07[b] (71)
0.005	1.33 ± 0.06[c] (75)	1.19 ± 0.05[bc'] (67)	0.96 ± 0.04[ab'] (54)
0.010	1.04 ± 0.06[ab'] (57)	0.93 ± 0.04[ab'] (52)	0.65 ± 0.03[aa'] (37)

Values are expressed as the mean ± SEM of six replicates; the values in the parentheses indicate the percentage change with control value taken as 100 per cent. a, b and c Significantly different $p < 0.001$, $p < 0.01$ and $p < 0.05$ respectively compared with control; a', b' and c' Significantly different $p < 0.001$, $p < 0.01$ and $p < 0.05$ respectively from corresponding treated group when t-test was applied. Analysis of variance showed that the response to the JHA was dose-dependent $p < 0.001$.

Total protein levels in eggs, obtained from the crosses between treated males (*i.e.* emerged from treated culture) and normal females, were reduced to 80 per cent (1.43 ± 0.08 µg/egg), 75 per cent (1.33 ± 0.06 µg/egg) and 57 per cent (1.04 ± 0.05 µg/egg) of the control value following exposure of 0.001, 0.005, and 0.010 ppm concentrations of fenoxycarb respectively.

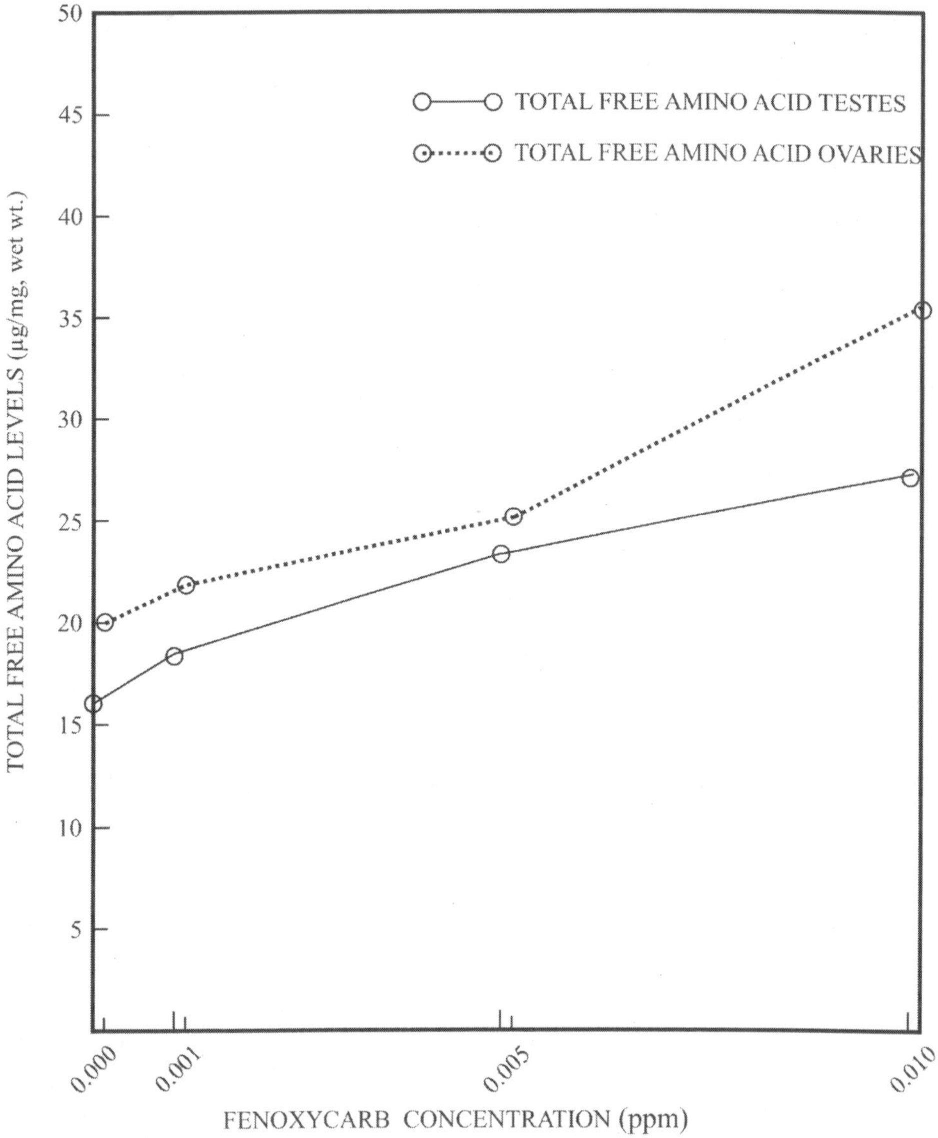

Figure 26: Graphic Representation of Effect of Fenoxycarb on the Total Free Amino Acid Levels in Testes and Ovaries of the Adult Rice Moth, *Corcyra cephalonica*.

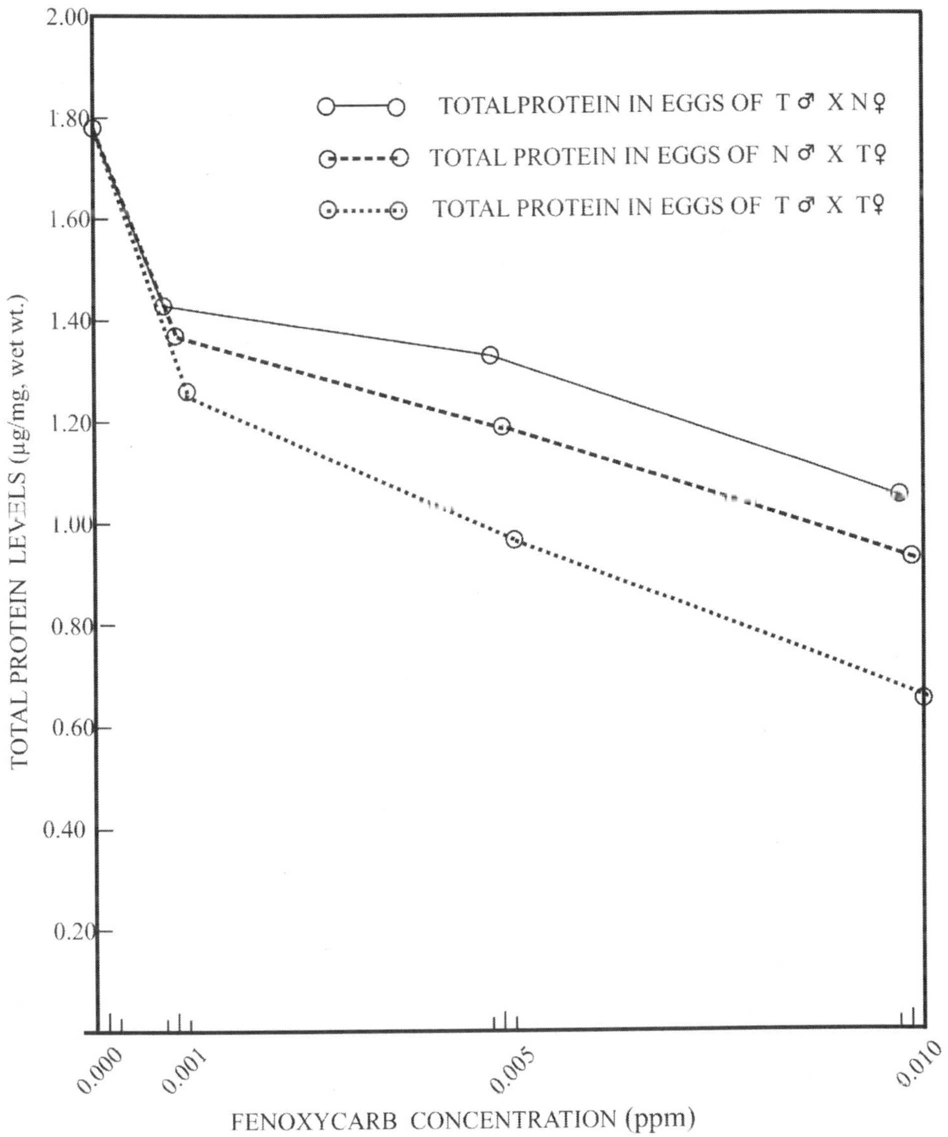

Figure 27: Graphic Representation of Effect of Fenoxycarb on the Total Protein, Levels in Eggs Obtained from different Crosses of the Adult Rice Moth, *Corcyra cephalonica* Treated with Sublethal Concentrations of Fenoxycarb.

The total protein levels in the eggs, collected from the crosses between normal males and treated females (*i.e.* emerged from treated culture), were recorded to be 77 per cent (1.37 ± 0.06 μg/egg), 67 per cent (1.19 ± 0.05 μg/egg) and 52 per cent (0.93 ± 0.04 μg/egg) of the control value following treatement with 0.001, 0.005 and 0.010 ppm concentrations of fenoxycarb respectively.

The total protein content in the eggs, collected from the crosses between sexes both emerged from treated cultures was observed to be 71 per cent (1.26 ± 0.07 μg/egg), 54 per cent (0.96 ± 0.04 μg/egg) and 37 per cent (0.65 ± 0.03 μg/egg) of the control value following exposure of 0.001, 0.005 and 0.010 ppm concentrations of fenoxycarb respectively.

Total Free Amino Acids Level in the Tests (Table 14, Figure 26)

Sub lethal concentrations of fenoxycarb caused a significantly dose-dependent ($p < 0.001$) enhancement in the levels of total free amino acid in the testes of rice moth, *C. cephalonica.* In case of control insects, the total free amino acids content in the testes was 16.07 ± 0.41 μg/mg. The maximum enhancement in total free amino acids level in testes (169 per cent of the control value) was observed in moths emerged from treated culture of 0.010 ppm concentration of fenoxycarb. Total free amino acid levels in testes were enhanced to 114 per cent (18.32 ± 0.47 μg/mg), 145 per cent (23.30 ± 0.95 μg/mg) and 169 per cent (27.16 ± 1.09 μg/mg) of the control value following exposure of 0.001, 0.005 and 0.010 ppm concentrations of fenoxycarb respectively.

Total Free Amino Acids Level in the Ovaries (Table 14, Figure 26)

Fenoxycarb at its sub lethal concentrations caused a significantly dose-dependent ($p < 0.001$) enhancement in the levels of total free amino acid in the ovaries of rice moth, *C. cephalonica.* The total free amino acids level, in the ovaries of control insect, was 20.02 ± 0.36 μg/mg. The maximum increase in total free amino acids level in ovaries (177 per cent of the control value) was observed in adults emerged from treated culture with 0.010 ppm concentration of fenoxycarb. Total free amino acid levels, in ovaries were increased to 109 per cent (21.82 ± 0.86 μg/mg), 125 per cent (25.03 ± 1.09 μg/mg) and 177 per cent (35.44 ± 1.31 μg/mg) of the control value following exposure of 0.001, 0.005 and 0.010 ppm concentrations of fenoxycarb respectively.

Total Free Amino Acids Level in the Eggs (Table 16, Figure 28)

Sub lethal concentrations of fenoxycarb caused a significantly dose-dependent ($p < 0.001$) enhancement in total free amino acids level in the eggs, collected from different crosses of *C. cephalonica.* The total free amino acids level in the eggs, obtained from control cross (normal male and normal female), was recorded to be 0.98 ± 0.06 μg/egg. The maximum increase in total free amino acids level in the eggs, collected from the crosses of treated male and normal female, normal male and treated female, and treated male and treated female, was recorded to be 120, 122 and 127 per cent of the control value respectively following treatment with 0.010 ppm concentration of fenoxycarb.

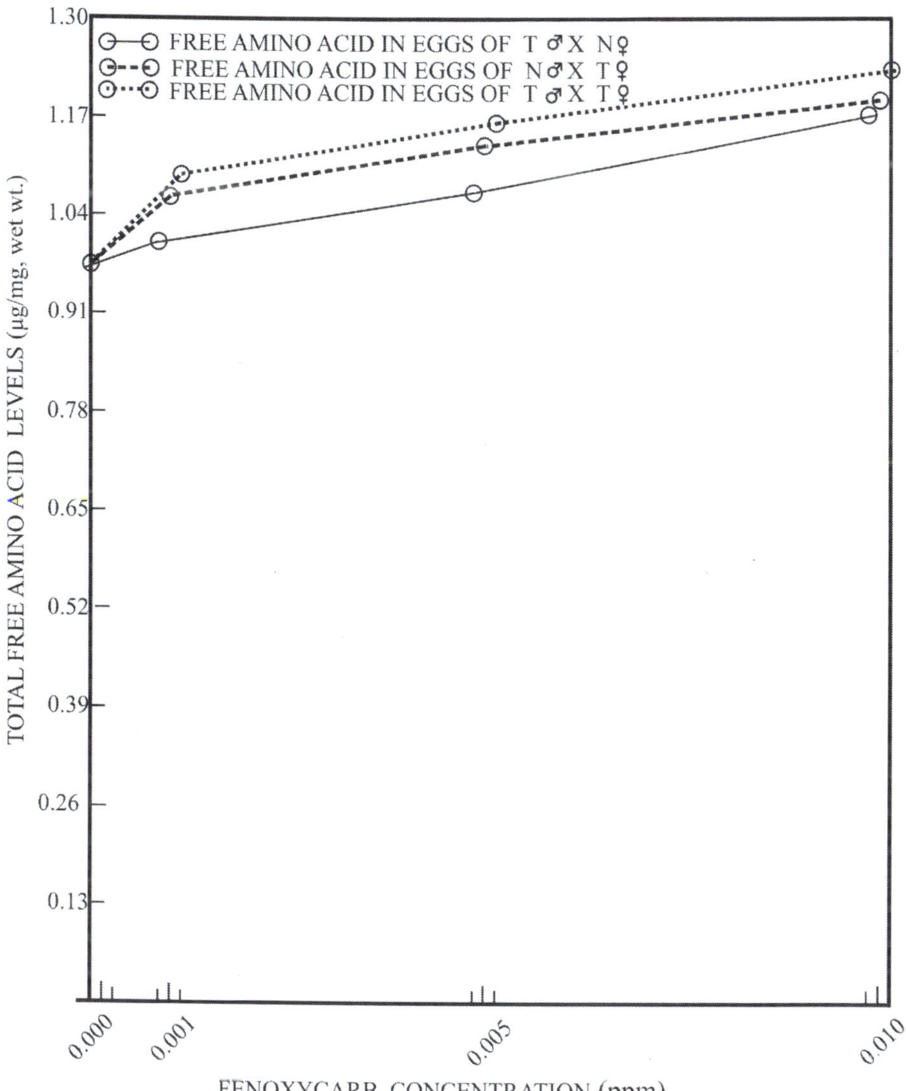

Figure 28: Graphic Representation of Effect of Fenoxycarb on the Total Free Amino Acid, Levels in Eggs Obtained from different Crosses of the Adult Rice Moth, *Corcyra cephalonica* Treated with Sublethal Concentrations of Fenoxycarb.

Table 16: Effect of Fenoxycarb on the Total Free Amino Acid Level (µg/egg) in the Eggs of *C. cephalonica*#

Fenoxycarb Concentration (ppm)	Treated Male x Normal Female	Normal Male x Treated Female	Treated Male x Treated Female
0.00 (control)	0.98 ± 0.06 (100)		
0.001	1.01 ± 0.08 (103)	1.07 ± 0.05 (109)	1.10 ± 0.05 (112)
0.005	1.08 ± 0.06 (110)	1.14 ± 0.04 (116)	1.17 ± 0.06[c] (119)
0.010	1.18 ± 0.06[b] (120)	1.20 ± 0.05[c] (122)	1.24 ± 0.04[b] (127)

Values are expressed as the mean ± SEM of six replicates; the values in the parentheses indicate the percentage change with control value taken as 100 per cent. b and c Significantly different $p < 0.01$ and $p < 0.05$ respectively compared with control when t-test was applied. Analysis of variance showed that the response to the JHA was dose-dependent $p < 0.001$.

Total free amino acid levels in eggs, obtained from the crosses between treated males (*i.e.* emerged from treated cultures) and normal females, were increased to 103 per cent (1.01 ± 0.08 µg/egg), 110 per cent (1.08 ± 0.06 µg/egg) and 120 per cent (1.18 ± 0.06 µg/egg) of the control value following treatment with 0.001, 0.005, and 0.010 ppm concentrations of fenoxycarb respectively.

The total free amino acid levels in the eggs, collected from the crosses between normal males and treated females (*i.e.* emerged from treated cultures), were recorded to be 109 per cent (1.07 ± 0.05 µg/egg), 116 per cent (1.14 ± 0.04 µg/egg) and 122 per cent (1.20 ± 0.05 µg/egg) of the control value when treated with 0.001, 0.005 and 0.010 ppm concentrations of fenoxycarb respectively.

The total free amino acid contents in the eggs, obtained from the crosses between treated males and treated females (*i.e.* both sexes emerged from treated cultures) were observed to be 112 per cent (1.10 ± 0.05 µg/egg), 119 per cent (1.17 ± 0.06 µg/egg) and 127 per cent (1.24 ± 0.04 µg/egg) of the control value following exposure of 0.001, 0.005 and 0.010 ppm concentrations of fenoxycarb respectively.

DNA Level in the Tests (Table 17, Figure 29)

Sub lethal concentrations of fenoxycarb caused a significantly dose-dependent ($p < 0.01$) reduction in the levels of DNA in the testes of rice moth, *C. cephalonica*. In case of control insects, the DNA content in the testes was 32.62 ± 0.22 µg/mg. The maximum decrease in DNA level in testes (77 per cent of the control value) was observed in moths emerged from treated culture of 0.010 ppm concentration of fenoxycarb. DNA levels in testes, were reduced to 92 per cent (30.01 ± 0.98 µg/mg), 87 per cent (28.38 ± 0.81 µg/mg) and 77 per cent (25.12 ± 0.36 µg/mg) of the control value following exposure of 0.001, 0.005 and 0.010 ppm concentrations of fenoxycarb respectively.

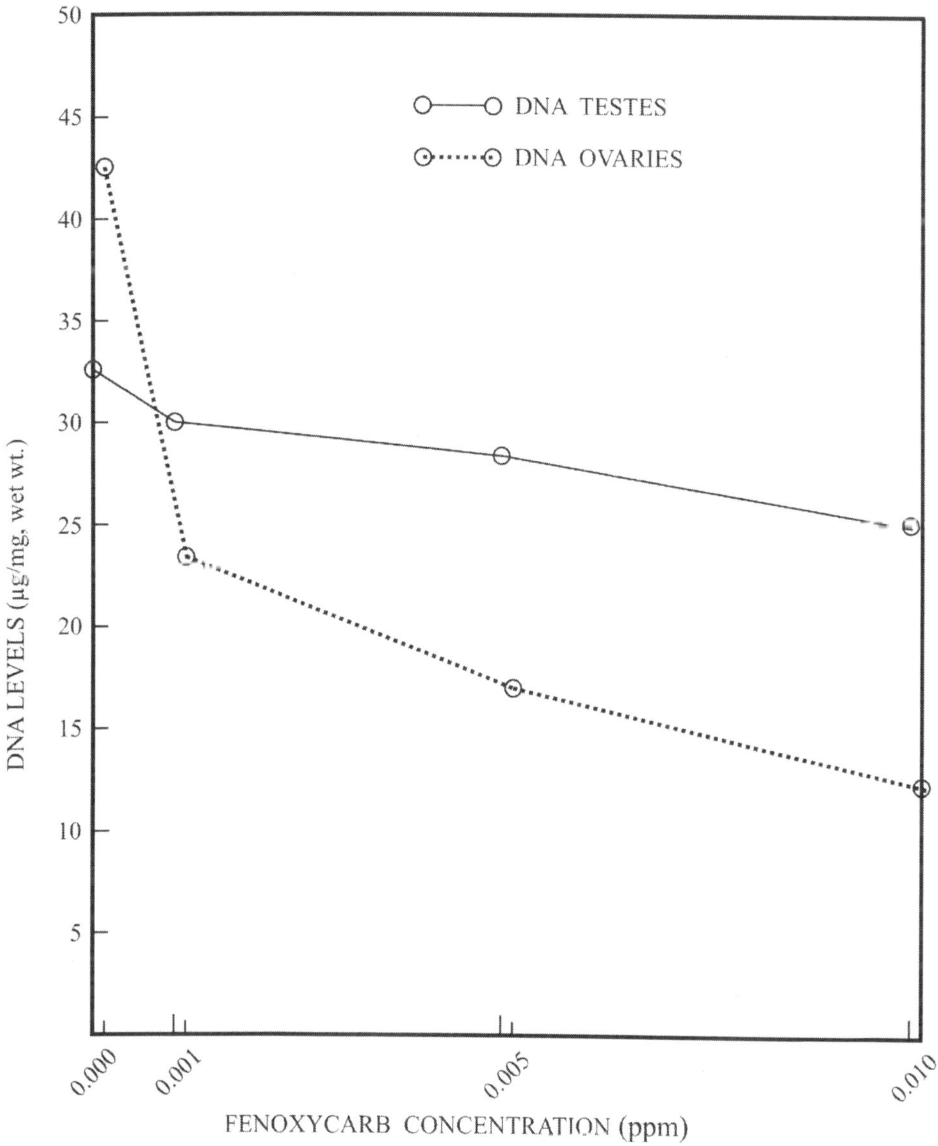

Figure 29: Graphic Representation of Changes in Deoxyribonucleic Acid Levels in Testes and Ovaries of the Adult Rice Moth, *Corcyra cephalonica* Treated with Sublethal Concentrations of Fenoxycarb.

Table 17: Changes in the DNA and RNA Levels in the Testes and Ovaries of the Rice Moth, *C. cephalonica* Treated with Fenoxycarb

Fenoxycarb Concentration (ppm)	*DNA[#] (µg/mg, wet wt.)*		*RNA[#] (µg/mg, wet wt.)*	
	Testes	*Ovaries*	*Testes*	*Ovaries*
Control	32.62 ± 0.22 (100)	42.55 ± 0.21 (100)	28.22 ± 0.42 (100)	31.84 ± 0.31 (100)
0.001	30.01 ± 0.98[c] (92)	23.38 ± 1.05[a] (55)	22.86 ± 0.48[a] (81)	23.88 ± 0.47[a] (75)
0.005	28.38 ± 0.81[a] (87)	17.00 ± 0.86[aa'] (40)	20.60 ± 0.49[ab'] (73)	21.01 ± 0.40[aa'] (66)
0.010	25.12 ± 0.36[ab'] (77)	12.33 ± 0.73[ab'] (29)	16.37 ± 0.40[aa'] (58)	15.92 ± 0.36[aa'] (50)

\# Values are expressed as the mean ± SEM of six replicates; the values in the parentheses indicate the percentage change with control values taken as 100 per cent. a and c Significantly different p< 0.001and p < 0.05 respectively compared with controls; a' and b' Significantly different p < 0.001 and p < 0.01 respectively from the corresponding treated group, when t-test was applied. Analysis of variance showed that the response to the JHA was dose-dependent p < 0.01.

DNA Level in the Ovaries (Table 17, Figure 29)

Fenoxycarb at its sub lethal concentrations caused a significantly dose-dependent (p < 0.01) reduction in the levels of DNA in the ovaries of rice moth, *C. cephalonica*. The DNA level in the ovaries of control insects, was 42.55 ± 0.21 µg/mg. The maximum decrease in DNA level in ovaries (29 per cent of the control value) was recorded in females emerged from treated culture of 0.010 ppm concentration of fenoxycarb. DNA levels in ovaries, were reduced to 55 per cent (23.38 ± 1.05 µg/mg), 40 per cent (17.00 ± 0.86 µg/mg) and 29 per cent (12.33 ± 0.73 µg/mg) of the control value following treatment with 0.001, 0.005 and 0.010 ppm concentrations of fenoxycarb respectively.

DNA Level in the Eggs (Table 18, Figure 31)

Exposure of sub lethal concentrations of fenoxycarb caused a significantly dose-dependent (p < 0.01) reduction in DNA levels in the eggs, collected from different crosses of *C. cephalonica*. The DNA content in the eggs, obtained from control cross (normal male and normal female), was recorded to be 4.27 ± 0.15 µg/egg. The maximum decrease in DNA level in the eggs, collected from the crosses of treated males and normal females, normal males and treated females, and treated males and treated females, was recorded to be 63, 53 and 39 per cent of the control value respectively following treatment with 0.010 ppm concentration of fenoxycarb.

DNA levels in eggs, obtained from the crosses between treated males (*i.e.* emerged from treated culture) and normal females, were reduced to 89 per cent (3.82 ± 0.12 µg/egg), 69 per cent (2.94 ± 0.146 µg/egg) and 63 per cent (2.68 ± 0.11 µg/egg) of the control value following exposure of 0.001, 0.005, and 0.010 ppm concentrations of fenoxycarb respectively.

Table 18: Effect of Fenoxycarb on the DNA Level (µg/egg) in the Eggs of *Corcyra cephalonica*#

Fenoxycarb Concentration (ppm)	*Treated Male x Normal Female*	*Normal Male x Treated Female*	*Treated Male x Treated Female*
0.00 (control)	4.27 ± 0.15 (100)	–	–
0.001	3.82 ± 0.12^{c} (89)	3.30 ± 0.14^{a} (77)	2.71 ± 0.10^{a} (63)
0.005	2.94 ± 0.14$^{aa'}$ (69)	2.85 ± 0.12$^{ac'}$ (67)	2.57 ± 0.15^{a} (60)
0.010	2.68 ± 0.11^{a} (63)	2.27 ± 0.15$^{ac'}$ (53)	1.68 ± 0.14$^{ab'}$ (39)

Values are expressed as the mean ± SEM of six replicates; the values in the parentheses indicate the percentage change with control value taken as 100 per cent. a, b and c Significantly different $p < 0.001$, $p < 0.01$ and $p < 0.05$ respectively compared with control; a' b' and c' Significantly different $p < 0.001$, $p < 0.01$ and $p < 0.05$ respectively from the corresponding treated group when t-test was applied. Analysis of variance showed that the response to the JHA was dose-dependent $p < 0.01$.

The DNA levels in the eggs, collected from the crosses between normal males and treated females (*i.e.* emerged from treated culture), were recorded to be 77 per cent (3.30 ± 0.14 µg/egg), 67 per cent (2.85 ± 0.12 µg/egg) and 53 per cent (2.27 ± 0.15 µg/egg) of the control value following treatement with 0.001, 0.005 and 0.010 ppm concentrations of fenoxycarb respectively.

The DNA contents in the eggs, collected from the crosses between sexes both emerged from treated cultures were observed to be 63 per cent (2.71 ± 0.10 µg/egg), 60 per cent (2.57 ± 0.15 µg/egg) and 39 per cent (1.68 ± 0.14 µg/egg) of the control value when treated with 0.001, 0.005 and 0.010 ppm concentrations of fenoxycarb respectively.

RNA Level in the Testes (Table 17, Figure 30)

A significantly dose-dependent ($p < 0.01$) reduction in the levels of RNA in the testes of rice moth, *C. cephalonica* was observed following exposure of sub lethal concentrations of fenoxycarb. In case of control insects, the RNA content in the testes was 28.22 ± 0.42 µg/mg. The maximum decrease in RNA level in testes (58 per cent of the control value) was observed in moths emerged from treated culture of 0.010 ppm concentration of fenoxycarb. The RNA levels in testes were reduced to 81 per cent (22.86 ± 0.48 µg/mg), 73 per cent (20.60 ± 0.49 µg/mg) and 58 per cent (16.37 ± 0.40 µg/mg) of the control value following exposure of 0.001, 0.005 and 0.010 ppm concentrations of fenoxycarb respectively.

RNA Level in the Ovaries (Table 17, Figure 30)

Fenoxycarb caused a significantly dose-dependent ($p < 0.01$) reduction in the levels of RNA in ovaries of rice moth, *C. cephalonica* when exposed at its sub lethal concentrations. The RNA level, in the ovaries of control insect, was 31.84 ± 0.31

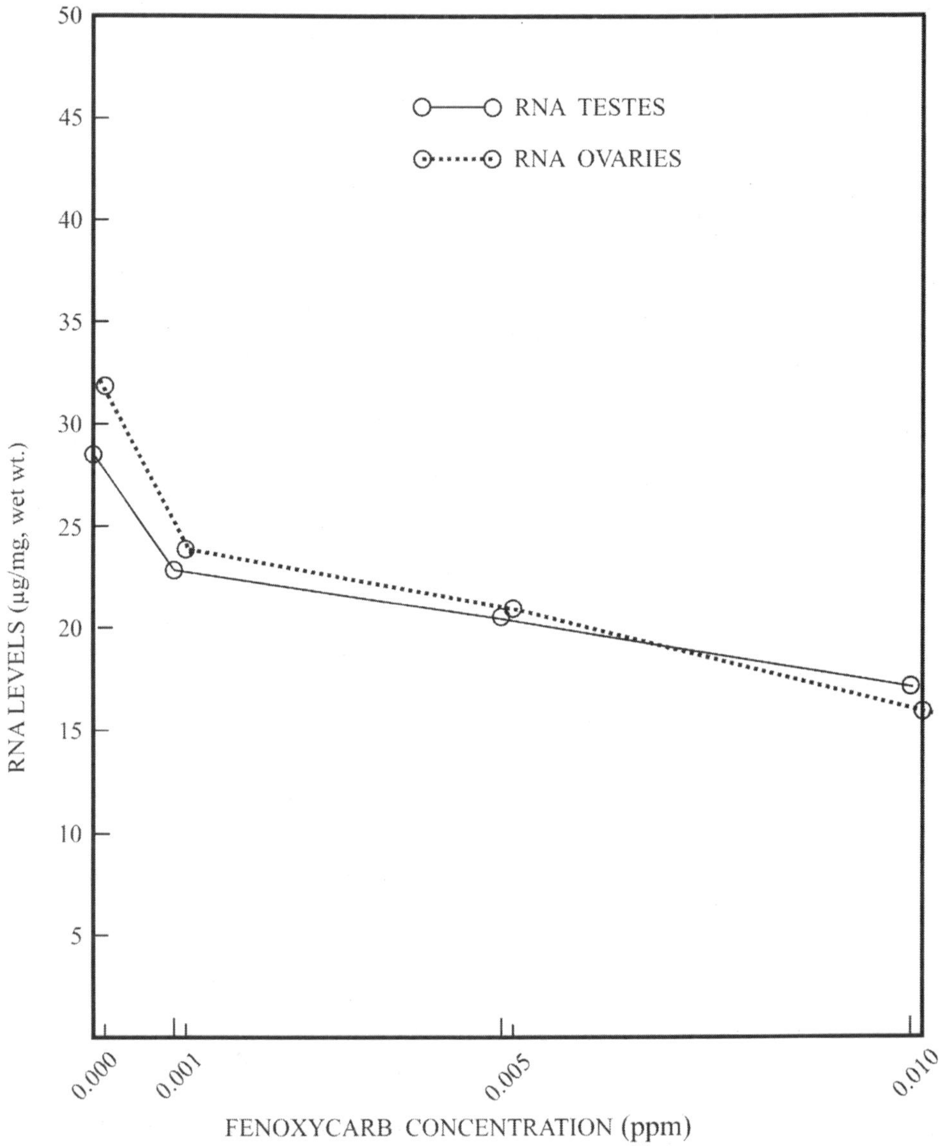

Figure 30: Graphic Representation of Changes in Ribonucleic Acid Levels in Testes and Ovaries of the Adult Rice Moth, *Corcyra cephalonica* Treated with Sublethal Concentrations of Fenoxycarb.

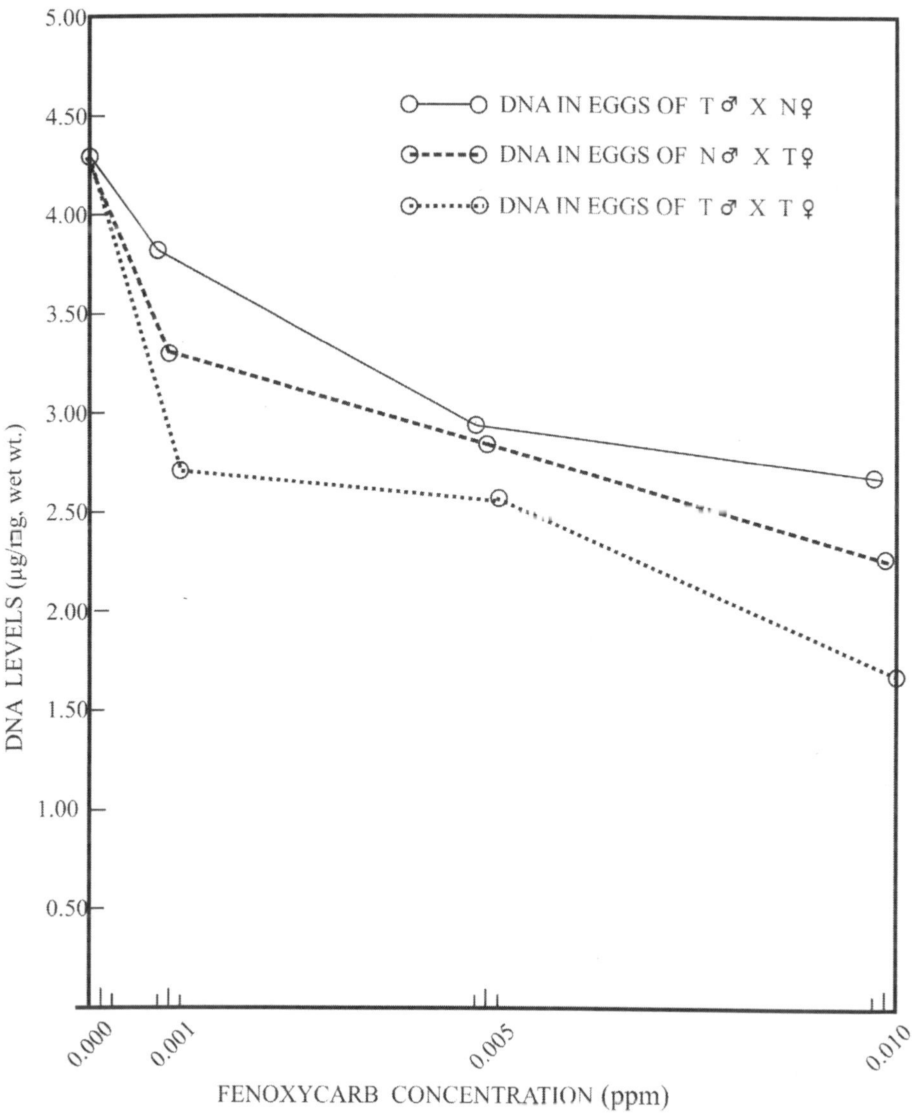

Figure 31: Graphic Representation of Effect of Fenoxycarb on the DNA, Levels in Eggs Obtained from different Crosses of the Rice Moth, *Corcyra cephalonica* Treated with Sublethal Concentrations of Fenoxycarb.

µg/mg. The maximum decrease in RNA level in ovaries (50 per cent of the control value) was recorded in females emerged from treated culture with 0.010 ppm concentration of fenoxycarb. The RNA levels, in ovaries were reduced to 75 per cent (23.88 ± 0.47 µg/mg), 66 per cent (21.01 ± 0.40 µg/mg) and 50 per cent (15.92 ± 0.36 µg/mg) of the control value following treatment with 0.001, 0.005 and 0.010 ppm concentrations of fenoxycarb respectively.

RNA Level in the Eggs (Table 19, Figure 32)

Exposure of sub lethal concentrations of fenoxycarb caused a significantly dose-dependent ($p < 0.01$) reduction in the RNA levels in the eggs, collected from different crosses of *C. cephalonica.* The RNA content in the eggs, obtained from control cross (normal male and normal female), was recorded to be 0.92 ± 0.06 µg/egg.

Table 19: Effect of Fenoxycarb on the RNA Level (µg/egg) in the Eggs of *C. cephalonica*[#]

Fenoxycarb Concentration (ppm)	*Treated Male x Normal Female*	*Normal Male x Treated Female*	*Treated Male x Treated Female*
0.00 (control)	0.92 ± 0.06 (100)		
0.001	0.82 ± 0.05 (91)	0.75 ± 0.07 (82)	0.70 ± 0.06[c] (76)
0.005	0.68 ± 0.04[c] (74)	0.62 ± 0.06[a] (67)	0.53 ± 0.05[a] (58)
0.010	0.54 ± 0.04[ac'] (59)	0.46 ± 0.04[a] (50)	0.40 ± 0.03[ac'] (43)

Values are expressed as the mean ± SEM of six replicates; the values in the parentheses indicate the percentage change with control values taken as 100 per cent. a, b and c Significantly different $p < 0.001$, $p < 0.01$ and $p < 0.05$ respectively compared with control; a' and c' Significantly different $p < 0.001$ and $p < 0.05$ respectively from the corresponding treated group when t-test was applied. Analysis of variance showed that the response to the JHA was dose-dependent $p < 0.01$.

The maximum decrease in RNA level in the eggs, collected from the crosses of treated males and normal females, normal males and treated females, and treated males and treated females, was recorded to be 59, 50 and 43 per cent of the control value respectively following treatment with 0.010 ppm concentration of fenoxycarb.

The RNA levels in eggs, obtained from the crosses between treated males (*i.e.* emerged from treated culture) and normal females, were reduced to 91 per cent (0.82 ± 0.05 µg/egg), 74 per cent (0.68 ± 0.04 µg/egg) and 59 per cent (0.54 ± 0.04 µg/egg) of the control value following exposure of 0.001, 0.005, and 0.010 ppm concentrations of fenoxycarb respectively.

The RNA levels in the eggs, collected from the crosses between normal males and treated females (*i.e.* emerged from treated culture), were recorded to be 82 per cent (0.75 ± 0.07 µg/egg), 67 per cent (0.62 ± 0.06 µg/egg) and 50 per cent (0.46 ± 0.04 µg/egg) of the control value following treatment with 0.001, 0.005 and 0.010 ppm concentrations of fenoxycarb respectively.

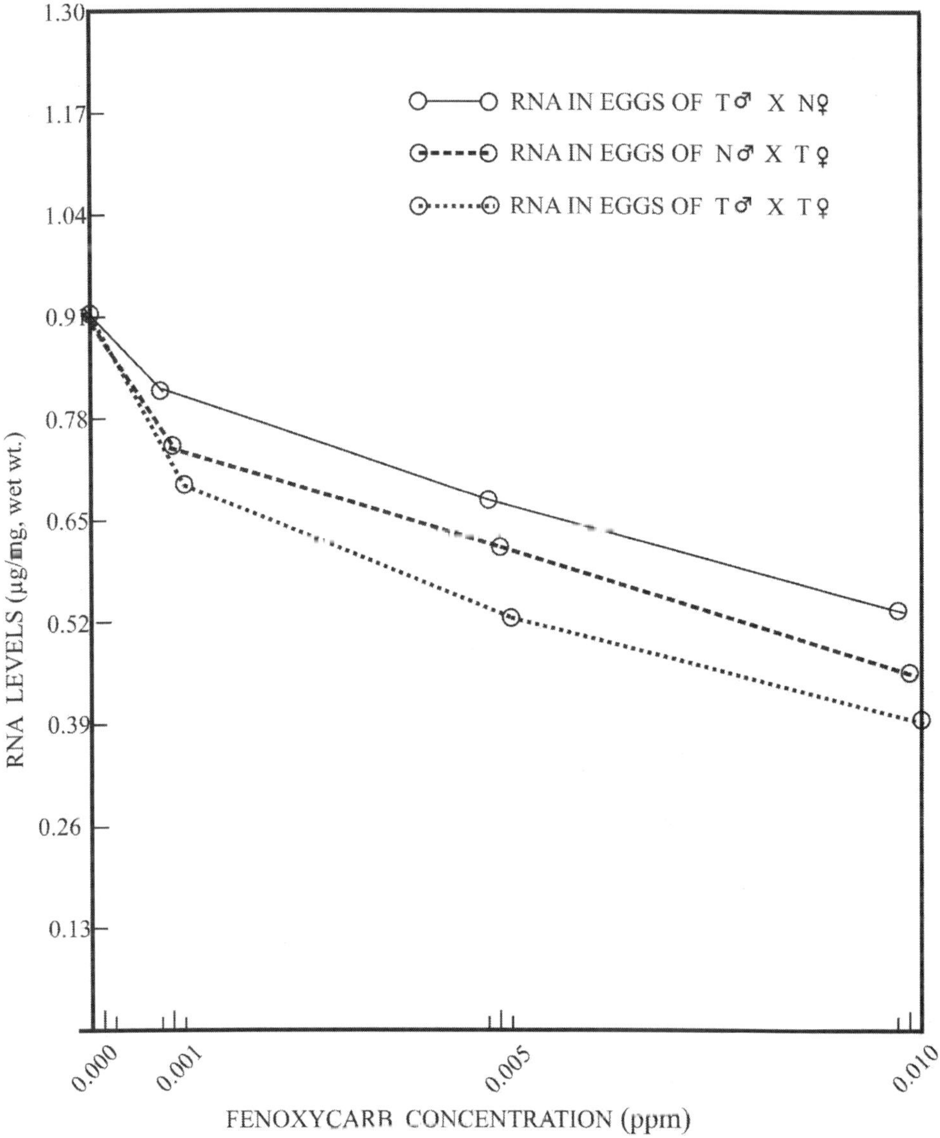

Figure 32: Graphic Representation of Effect of Fenoxycarb on the RNA, Levels in Eggs Obtained from different Crosses of the Rice Moth, *Corcyra cephalonica* Treated with Sublethal Concentrations of Fenoxycarb.

The RNA content in the eggs, obtained from the crosses of treated males and treated females (*i.e.* both sexes emerged from treated culture), were observed to be 76 per cent (0.70 ± 0.06 μg/egg), 58 per cent (0.53 ± 0.05 μg/egg) and 43 per cent (0.40 ± 0.03 μg/egg) of the control value following exposure of 0.001, 0.005 and 0.010 ppm concentrations of fenoxycarb respectively.

Glycogen Level in the Testes (Table 20, Figure 33)

Sub lethal concentrations of fenoxycarb caused a significantly dose-dependent ($p < 0.05$) reduction in the level of total protein in the testes of rice moth. In case of control insect, the glycogen level in the testes was 2.31 ± 0.06 μg/mg. The maximum decrease in glycogen level in testes (42 per cent of the control value) was observed in moths emerged from treated culture of 0.010 ppm concentration of fenoxycarb. The glycogen levels in testes, were reduced to 84 per cent (1.94 ± 0.01 μg/mg), 67 per cent (1.55 ± 0.03 μg/mg) and 42 per cent (0.97 ± 0.02 μg/mg) of the control value following exposure of 0.001, 0.005 and 0.010 ppm concentrations of fenoxycarb respectively.

Table 20: Changes in the Levels of Glycogen in the Testes and Ovaries of the Rice Moth, *C. cephalonica* Treated with Fenoxycarb

Fenoxycarb concentration (ppm)	*Glycogen# (μg/mg, wet wt.)*	
	Testes	*Ovaries*
Control	2.31 ± 0.06 (100)	2.14 ± 0.07 (100)
0.001	1.94 ± 0.01[a] (84)	1.24 ± 0.01[a] (58)
0.005	1.55 ± 0.03[aa'] (67)	0.96 ± 0.02[aa'] (45)
0.010	0.97 ± 0.02[aa'] (42)	0.39 ± 0.01[aa'] (18)

Values are expressed as the mean ± SEM of six replicates; the values in the parentheses indicate the percentage change with control values taken as 100 per cent. a Significantly different $p < 0.001$ compared with controls; a' Significantly different $p < 0.001$ from the corresponding treated group when t-test was applied. Analysis of variance showed that the response to the JHA was dose-dependent $p < 0.05$.

Glycogen Level in the Ovaries (Table 20, Figure 33)

Sub lethal concentrations of fenoxycarb caused a significantly dose-dependent ($p < 0.05$) reduction in the level glycogen in the ovaries of *C. cephalonica*. The glycogen level, in the ovaries of control insect, was 2.14 ± 0.07 μg/mg. Fenoxycarb at its 0.010 ppm concentration caused maximum reduction in glycogen level in ovaries (36 per cent of the control value) of females emerged from treated culture. The glycogen levels in ovaries, were reduced to 58 per cent (1.24 ± 0.01 μg/mg), 45 per cent (0.96 ± 0.02 μg/mg) and 18 per cent (0.39 ± 0.01 μg/mg) of the control value when treated with 0.001, 0.005 and 0.010 ppm concentrations of fenoxycarb respectively.

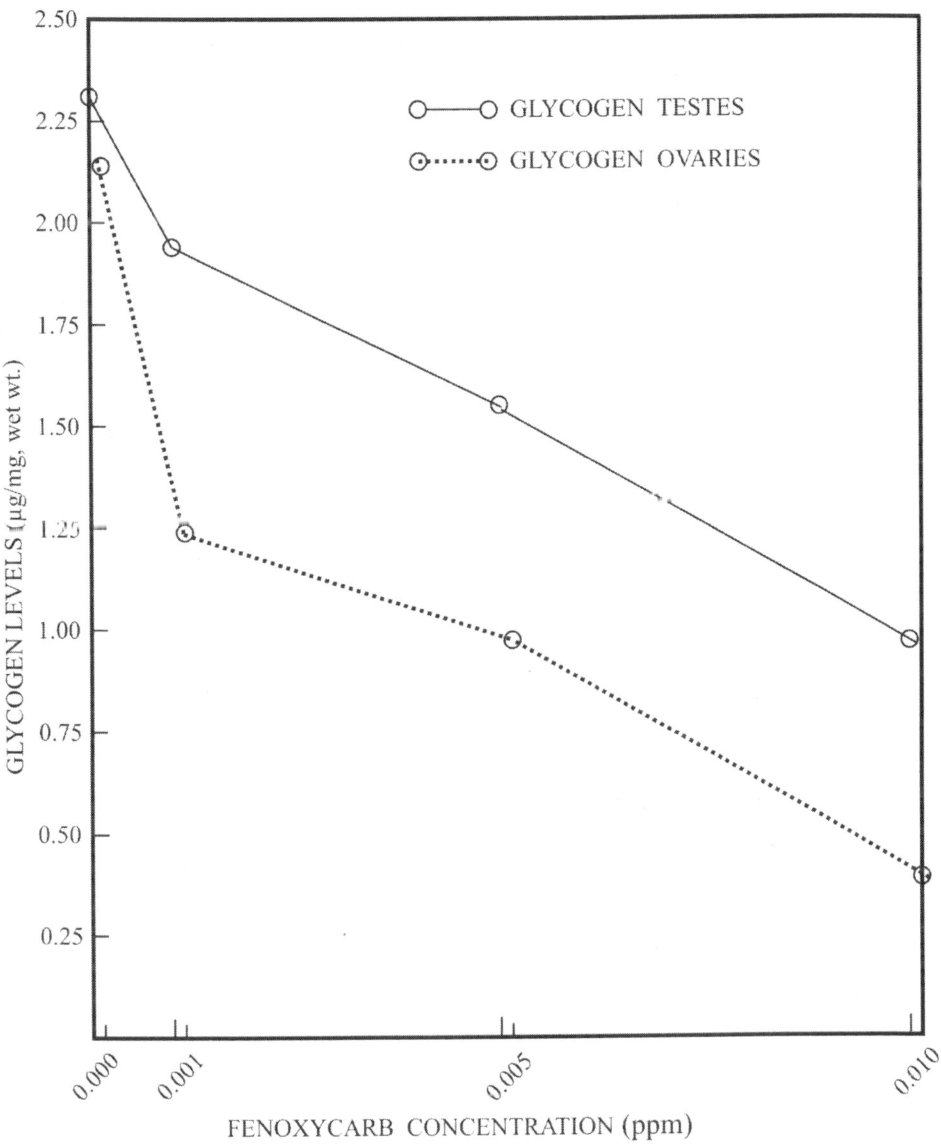

Figure 33: Graphic Representation of Changes in Glycogen Levels in Testes and Ovaries of the Adult Rice Moth, *Corcyra cephalonica* Treated with Sublethal Concentrations of Fenoxycarb.

Glycogen Level in the Eggs (Table 21, Figure 34)

Exposure of sub lethal concentrations of fenoxycarb caused a significantly dose-dependent ($p < 0.05$) reduction in glycogen level in the eggs, collected from different crosses of *C. cephalonica.* The glycogen content in the eggs, obtained from control cross (normal male and normal female), was recorded to be 0.93 ± 0.06 µg/egg. The maximum decrease in glycogen level in the eggs, collected from the crosses of treated male and normal female, normal male and treated female, and treated male and treated female, was recorded to be 63, 53 and 42 per cent of the control value respectively following treatment with 0.010 ppm concentration of fenoxycarb.

Table 21: Effect of Fenoxycarb on the Glycogen Level (µg/egg) in the Eggs of *Corcyra cephalonica*[#]

Fenoxycarb Concentration (ppm)	*Treated Male x Normal Female*	*Normal Male x Treated Female*	*Treated Male x Treated Female*
0.00 (control)	0.93 ± 0.06 (100)	–	–
0.001	0.74 ± 0.05[c] (80)	0.69 ± 0.04[b] (75)	0.65 ± 0.05[b] (69)
0.005	0.67 ± 0.04[b] (72)	0.63 ± 0.03[b] (67)	0.53± 0.04[a] (55)
0.010	0.58 ± 0.03[a] (63)	0.51 ± 0.02[ab'] (53)	0.38 ± 0.03[ac'] (42)

Values are expressed as the mean ± SEM of six replicates; the values in the parentheses indicate the percentage change with control value taken as 100 per cent. a, b and c Significantly different $p < 0.001$, $p < 0.01$ and $p < 0.05$ respectively compared with control; b' and c' significantly different $p < 0.01$ and $p < 0.05$ respectively from the corresponding treated group when t-test was applied. Analysis of variance showed that the response to the JHA was dose-dependent $p < 0.05$.

Glycogen levels in eggs, obtained from the crosses between treated males (*i.e.* emerged from treated culture) and normal females, were reduced to 80 per cent (0.74 ± 0.05 µg/egg), 72 per cent (0.67 ± 0.04 µg/egg) and 63 per cent (0.58 ± 0.03 µg/egg) of the control value following exposure of 0.001, 0.005, and 0.010 ppm concentrations of fenoxycarb respectively.

The glycogen levels in the eggs, collected from the crosses between normal males and treated females (*i.e.* emerged from treated culture), were recorded to be 75 per cent (0.69 ± 0.04 µg/egg), 67 per cent (0.63 ± 0.03 µg/egg) and 53 per cent (0.51 ± 0.02 µg/egg) of the control value following treatment with 0.001, 0.005 and 0.010 ppm concentrations of fenoxycarb respectively.

The glycogen content in the eggs, collected from the crosses between treated males and treated females were found to be 69 per cent (0.65 ± 0.05 µg/egg), 55 per cent (0.53 ± 0.04 µg/egg) and 42 per cent (0.38 ± 0.03 µg/egg) of the control value following exposure of 0.001, 0.005 and 0.010 ppm concentrations of fenoxycarb respectively.

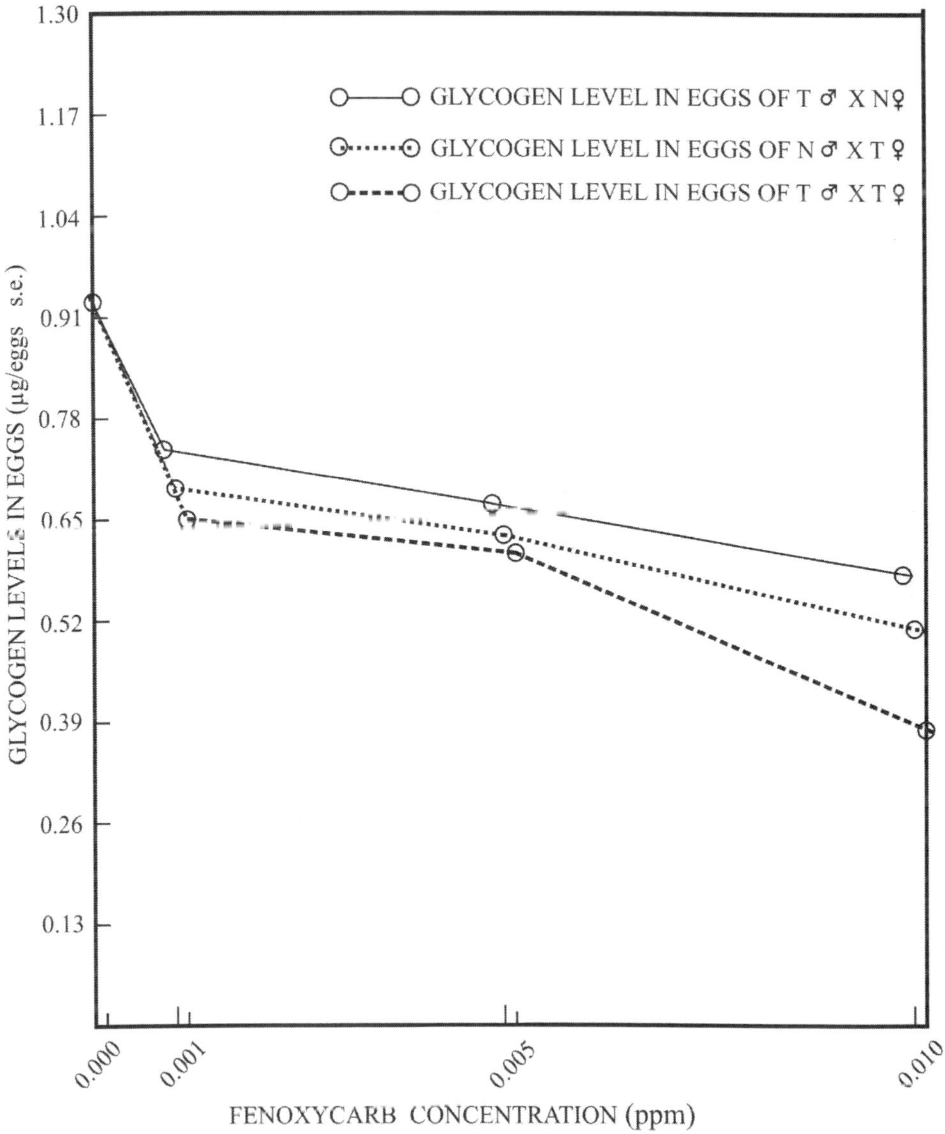

Figure 34: Graphic Representation of Effect of Fenoxycarb on the Glycogen Levels in Eggs Obtained from different Crosses of the Rice Moth, *Corcyra cephalonica*.

Activity of Acid Phosphatase in the Testes (Table 22, Figure 35)

Sub lethal concentrations of fenoxycarb caused a significantly dose-dependent ($p < 0.05$) enhancement in the activity of acid phosphatase in the testes of *C. cephalonica.*

Table 22: Changes in the Acid and Alkaline Phosphatase Activity in the Testes and Ovaries of the Rice Moth, *C. cephalonica* Treated with Fenoxycarb

Fenoxycarb Concentration (ppm)	*Acid Phosphatase#*		*Alkaline Phosphatase#*	
	Testes	*Ovaries*	*Testes*	*Ovaries*
Control	18.22 ± 0.24 (100)	21.02 ± 0.25 (100)	36.42 ± 0.29 (100)	44.00 ± 0.27 (100)
0.001	22.78 ± 0.95[a] (125)	24.17 ± 0.87[b] (115)	32.05 ± 0.84[a] (88)	40.92 ± 1.03[c] (93)
0.005	28.61 ± 0.69[aa'] (157)	30.48 ± 1.04[aa'] (145)	25.49 ± 0.89[aa'] (70)	36.96 ± 1.10[ac'] (84)
0.010	38.26 ± 1.09[aa'] (210)	37.42 ± 1.09[ab'] (178)	17.48 ± 0.38[aa'] (48)	28.60 ± 0.98[aa'] (65)

The activities are given as μ moles substrate hydrolyzed/30 minutes/mg protein and expressed as mean ± SEM of six replicates; the values in the parentheses are the percentage change with control values taken as 100 per cent. a, b and c Significantly different $p < 0.001$, $p < 0.01$ and $p < 0.05$ respectively compared with controls; a' b' and c' Significantly different $p < 0.001$, $p < 0.01$ and $p < 0.05$ respectively from the corresponding treated group when t-test was applied. Analysis of variance showed that the response to the JHA was dose-dependent $p < 0.05$.

Incase of control insects, the activity of acid phosphatase in the testes was 18.22 ± 0.24 μ moles substrate hydrolyzed/30 min/mg protein. The maximum enhancement in activity of acid phosphatase in testes (210 per cent of the control value) was observed in moths emerged from treated culture of 0.010 ppm concentration of fenoxycarb. The activities of acid phosphatase in the testes were enhanced to 125 per cent (22.78 ± 0.95), 157 per cent (28.61 ± 0.69) and 210 per cent (38.26 ± 1.09) μ moles substrate hydrolyzed/30 min/mg protein, of the control value following exposure of 0.001, 0.005 and 0.010 ppm concentrations of fenoxycarb respectively.

Activity of Acid Phosphatase in the Ovaries (Table 22, Figure 35)

Fenoxycarb at its sub lethal concentrations caused a significantly dose-dependent ($p < 0.05$) enhancement in the activity of acid phosphatase in the ovaries of rice moth, *C. cephalonica.* The activity of acid phosphatase, in the ovaries of control insect, was 21.02 ± 0.25 09 μ moles substrate hydrolyzed/30 min/mg protein. The maximum increase in activity of acid phosphatase in ovaries (178 per cent of the control value) was observed in females emerged from treated culture with 0.010 ppm concentration of fenoxycarb. The activities of acid phosphatase in ovaries, were increased to 115 per cent (24.17 ± 0.87), 145 per cent (30.48 ± 1.04) and 178 per cent (37.42 ± 1.09) μ moles substrate hydrolyzed/30 min/mg protein of the

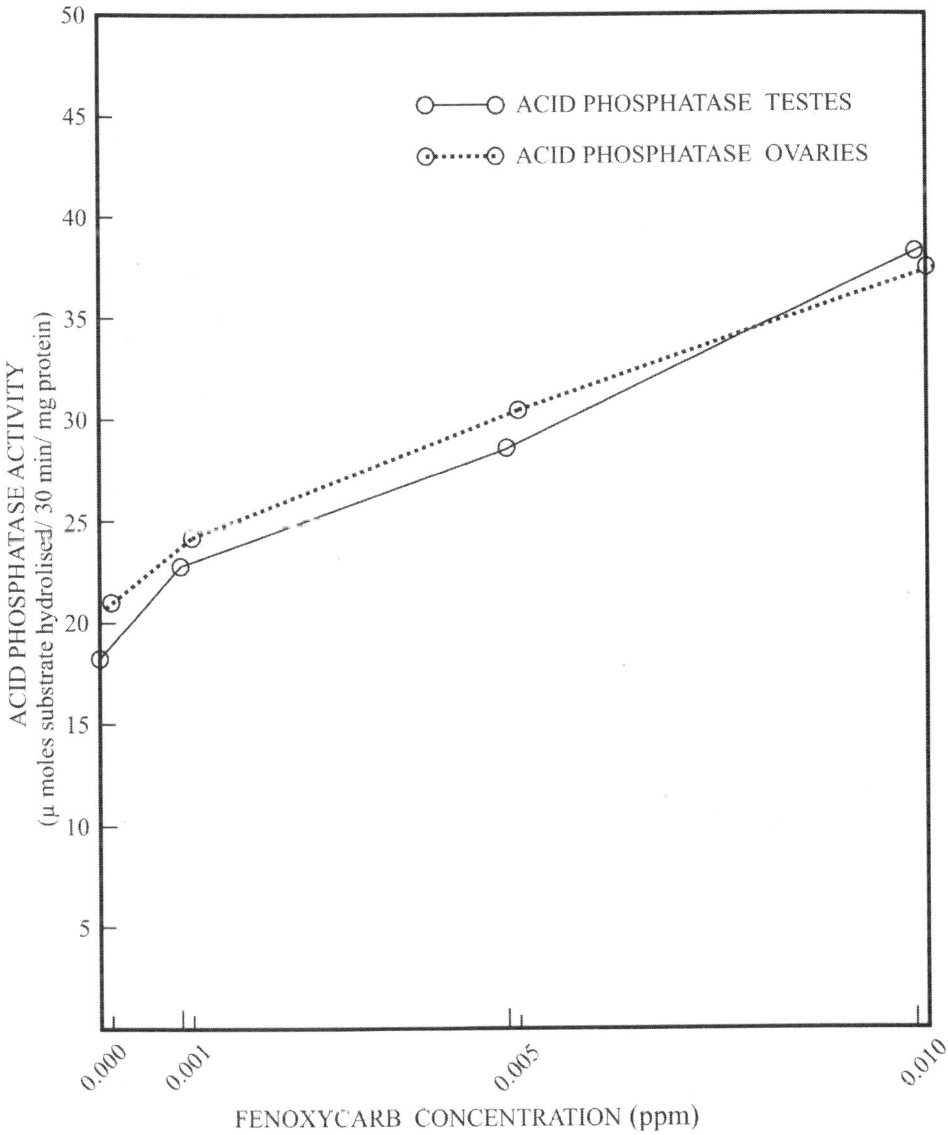

Figure 35: Graphic Representation of Changes in Acid Phosphatase Activities in Testes and Ovaries of the Adult Rice Moth, *Corcyra cephalonica* Treated with Sublethal Concentrations of Fenoxycarb.

control value following exposure of 0.001, 0.005 and 0.010 ppm concentrations of fenoxycarb respectively.

Activity of Acid Phosphatase in the Eggs (Table 23, Figure 37)

Sub lethal concentrations of fenoxycarb caused a significantly dose-dependent ($p < 0.05$) enhancement in the activity of acid phosphatase in the eggs, collected from differents crosses of *C. cephalonica.* The activity of acid phosphatase in the eggs, obtained from control cross (normal male and normal female), was recorded to be 14.75 ± 1.14 μ moles substrate hydrolyzed/30 minutes/mg protein. The maximum increase in the activity of acid phosphatase in the eggs, collected from the crosses of treated males and normal females, normal males and treated females, and treated males and treated females, was recorded to be 171, 189 and 196 per cent of the control value respectively following treatment with 0.010 ppm concentration of fenoxycarb.

Table 23: Effect of Fenoxycarb on the Activity of Acid Phosphatase in the Eggs of *C. cephalonica*[#]

Fenoxycarb Concentration (ppm)	*Treated Male x Normal Female*	*Normal Male x Treated Female*	*Treated Male x Treated Female*
0.00 (control)	14.75 ± 1.14 (100)	—	—
0.001	19.51 ± 0.99[a] (132)	21.95 ± 1.65[b] (149)	23.35 ± 1.52[b] (158)
0.005	22.95 ± 1.68[a] (156)	25.02 ± 1.44[a] (170)	26.02 ± 1.78[a] (176)
0.010	25.26 ± 1.50[a] (171)	27.88 ± 1.58[a] (189)	28.92 ± 1.94[a] (196)

The activities are given as μ moles substrate hydrolyzed/30 minutes/mg protein and expressed as mean ± SEM of six replicates; the values in the parentheses are the percentage change with control value taken as 100 per cent. a and b Significantly different $p < 0.001$, $p < 0.01$ and $p < 0.05$ respectively compared with control. Analysis of variance showed that the response to the JHA was dose-dependent $p < 0.05$.

The activities of acid phosphatase in eggs, obtained from the crosses between treated males (*i.e.* emerged from treated culture) and normal females, were increased to 132 per cent (19.51 ± 0.99), 156 per cent (22.95 ± 1.68) and 171 per cent (25.26 ± 1.50) μ moles substrate hydrolyzed/30 minutes/mg protein of the control value following treatment with 0.001, 0.005, and 0.010 ppm concentrations of fenoxycarb respectively.

The activities of acid phosphatase in the eggs, collected from the crosses between normal males and treated females (*i.e.* emerged from treated culture), were recorded to be 149 per cent (21.95 ± 1.65), 170 per cent (25.02 ± 1.44) and 189 per cent (27.88 ± 1.58) μ moles substrate hydrolyzed/30 min/mg protein of the control value when treated with 0.001, 0.005 and 0.010 ppm concentrations of fenoxycarb respectively.

The activities of acid phosphatase in the eggs, obtained from the crosses between treated males and treated females (*i.e.* both sexes emerged from treated culture)

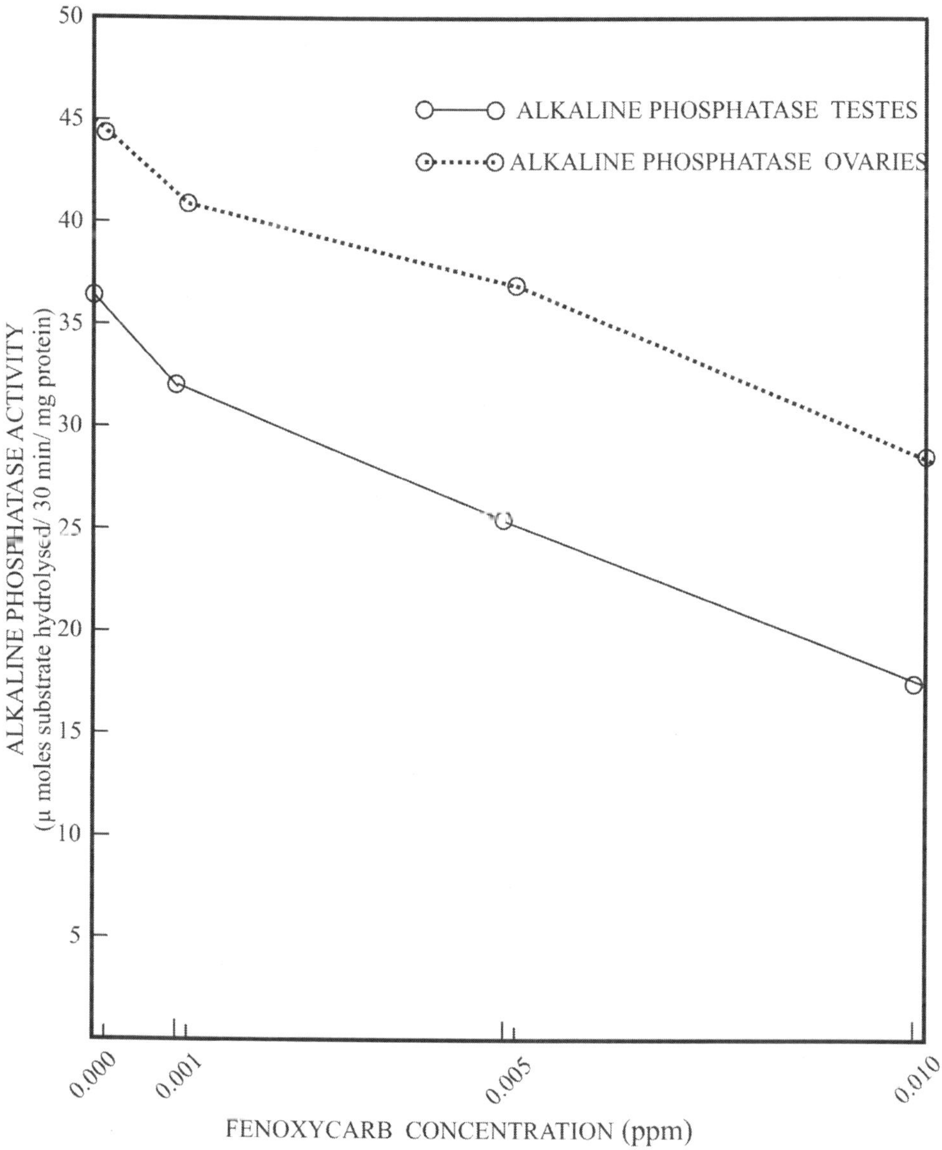

Figure 36: Graphic Representation of Changes in Alkaline Phosphatase Activities in Testes and Ovaries of the Adult Rice Moth, *Corcyra cephalonica* Treated with Sublethal Concentrations of Fenoxycarb.

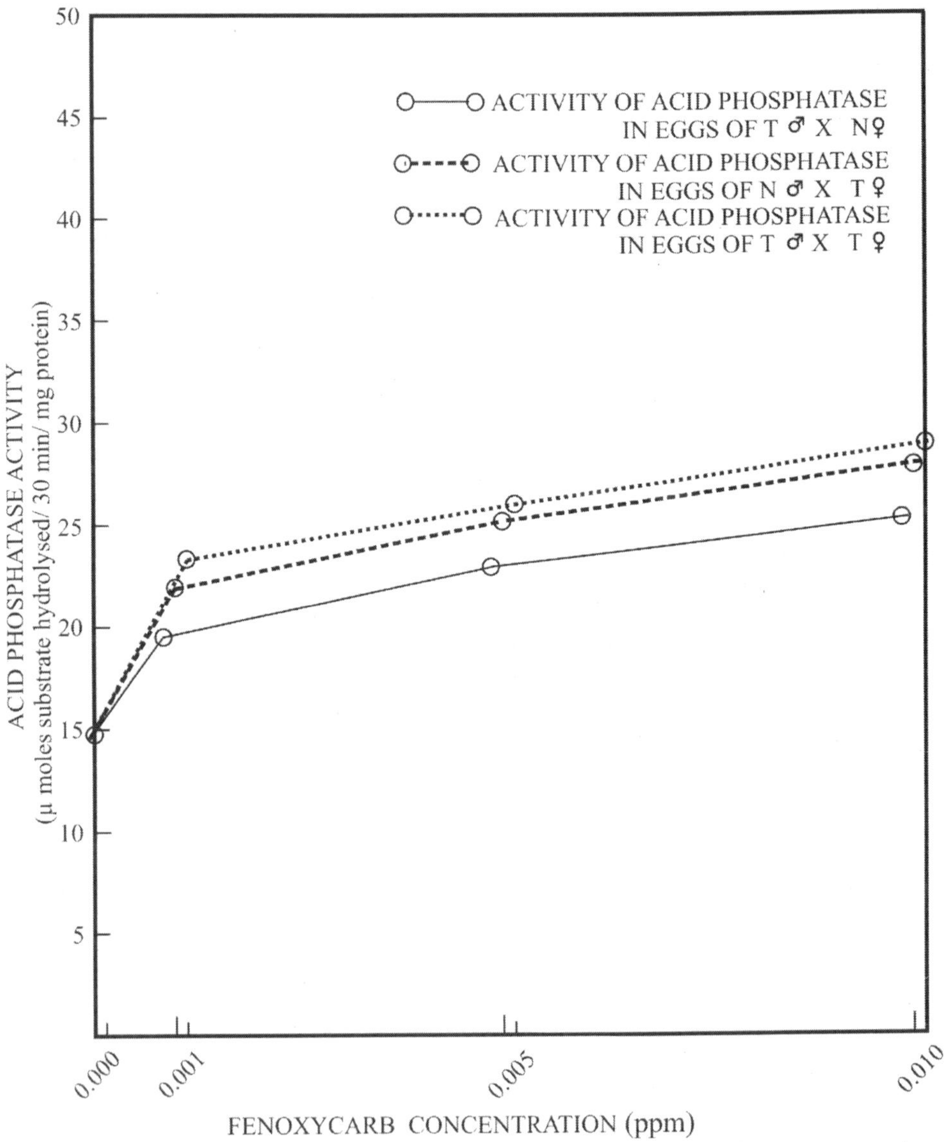

Figure 37: Graphic Representation of Changes in Acid Phosphatase Activities in Eggs Obtained from different Crosses of Rice Moth, *Corcyra cephalonica* Treated with Fenoxycarb.

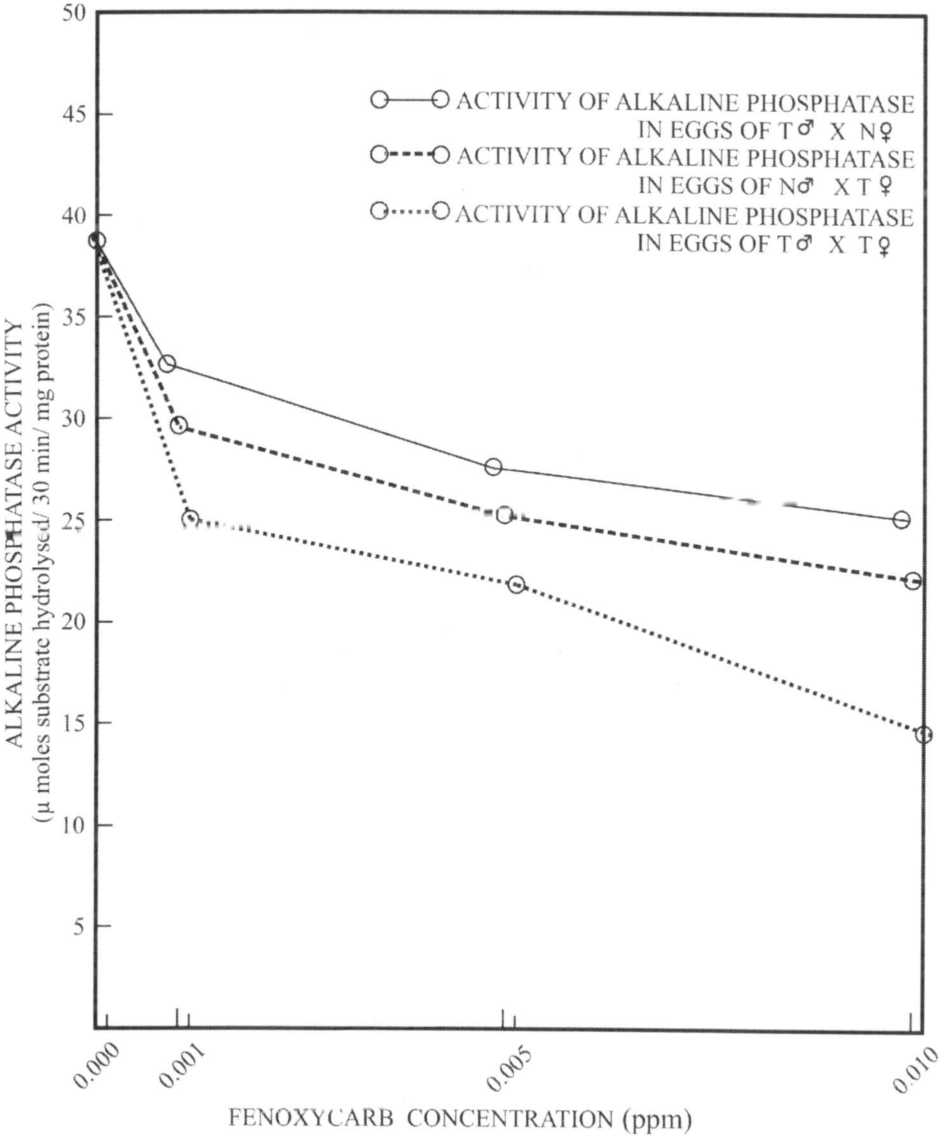

Figure 38: Graphic Representation of Changes in Alkaline Phosphatase Activities in Eggs Obtained from different Crosses of Rice Moth, *Corcyra cephalonica* Treated with Fenoxycarb.

were observed to be 158 per cent (23.35 ± 1.52), 176 per cent (26.02 ± 1.78) and 196 per cent (28.92 ± 1.94) µ moles substrate hydrolyzed/30 min/mg protein of the control value following exposure of 0.001, 0.005 and 0.010 ppm concentrations of fenoxycarb respectively.

Activity of Alkaline Phosphatase in the Testes (Table 22, Figure 36)

Sub lethal concentrations of fenoxycarb caused a significantly dose-dependent ($p < 0.05$) reduction in the activity of alkaline phosphatase in the testes of *C. cephalonica*. In case of control insect, the activity of alkaline phosphatase in the testes was 36.42 ± 0.29 µ moles substrate hydrolyzed/30 min/mg protein. The maximum reduction in the activity of alkaline phosphatase in testes (48 per cent of the control value) was observed in males emerged from treated culture of 0.010 ppm concentration of fenoxycarb. The activity of alkaline phosphatase in testes were reduced to 88 per cent (32.05 ± 0.84), 70 per cent (25.49 ± 0.89) and 48 per cent (17.48 ± 0.38) µ moles substrate hydrolyzed/30 min/mg protein of the control value following exposure of 0.001, 0.005 and 0.010 ppm concentrations of fenoxycarb respectively.

Activity of Alkaline Phosphatase in the Ovaries (Table 22, Figure 36)

Fenoxycarb at its sub lethal concentrations caused a significantly dose-dependent ($p < 0.05$) reduction in the activity of alkaline phosphatase in the ovaries of *C. cephalonica*. The activity of alkaline phosphatase in the ovaries of control insect, was 44.00 ± 0.27 µ moles substrate hydrolyzed/30 min/mg protein. The maximum decrease in activity of alkaline phosphatase in ovaries (65 per cent of the control value) was recorded in females emerged from treated culture with 0.010 ppm concentration of µ moles substrate hydrolyzed/30 min/mg protein of the control value following fenoxycarb. The activities of alkaline phosphatase in ovaries, were reduced to 93 per cent (40.92 ± 1.03), 84 per cent (36.96 ± 1.10) and 65 per cent (28.60 ± 0.98) treatment with 0.001, 0.005 and 0.010 ppm concentrations of fenoxycarb respectively.

Activity of Alkaline Phosphatase in the Eggs (Table 24, Figure 38)

Exposure of sub lethal concentrations of fenoxycarb caused a significantly dose-dependent ($p < 0.05$) reduction in the activity of alkaline phosphatase in the eggs, collected from different crosses of *C. cephalonica*. The activity of alkaline phosphatase in the eggs, obtained from control cross (normal males and normal females), was recorded to be 38.77 ± 2.14 µ moles substrate hydrolyzed/30 min/mg protein. The maximum decrease in the activity of alkaline phosphatase in the eggs, collected from the crosses of treated males and normal females, normal males and treated females, and treated males and treated females, was recorded to be 65, 56 and 39 per cent of the control value respectively following treatment with 0.010 ppm concentration of fenoxycarb.

The activities of alkaline phosphatase in eggs, obtained from the crosses between treated males (*i.e.* emerged from treated culture) and normal females, were reduced to 84 per cent (32.65 ± 1.85), 73 per cent (27.62 ± 1.75) and 65 per cent (25.21 ± 1.84) µ moles substrate hydrolyzed/30 min/mg protein of the control value following exposure of 0.001, 0.005, and 0.010 ppm concentrations of fenoxycarb respectively.

Table 24: Effect of Fenoxycarb on the Activity of Alkaline Phosphatase in the Eggs of *C. cephalonica*#

Fenoxycarb Concentration (ppm)	*Treated Male x Normal Female*	*Normal Male x Treated Female*	*Treated Male x Treated Female*
0.00 (control)	38.77 ± 2.14 (100)	–	–
0.001	32.65 ± 1.85 (84)	29.65 ± 1.74[b] (76)	25.02 ± 1.65[a] (63)
0.005	27.62 ± 1.75[b] (73)	25.11 ± 1.52[a] (65)	21.88 ± 1.32[a] (56)
0.010	25.21 ± 1.84[a] (65)	22.21 ± 1.66[a] (56)	14.96 ± 1.11[ab'] (39)

The activities are given as µ moles substrate hydrolyzed/30 minutes/mg protein and expressed as mean ± SEM of six replicates; the values in the parentheses are the percentage change with control value taken as 100 per cent. a and b Significantly different $p < 0.001$ and $p < 0.01$ respectively compared with control; b' Significantly different $p < 0.01$ from the corresponding treated group when t-test was applied. Analysis of variance showed that the response to the JHA was dose-dependent $p < 0.05$

The activities of alkaline phosphatase in the eggs, collected from the crosses between normal males and treated females (*i.e.* emerged from treated culture), were recorded to be 76 per cent (29.65 ± 1.74), 65 per cent (25.11 ± 1.52) and 56 per cent (22.21 ± 1.66) µ moles substrate hydrolyzed/30 min/mg protein of the control value following treatement with 0.001, 0.005 and 0.010 ppm concentrations of fenoxycarb respectively.

The activities of alkaline phosphatase in the eggs, collected from the crosses between sexes both emerged from treated culture, were observed to be 63 per cent (25.02 ± 1.65), 56 per cent (21.88 ± 1.32) and 39 per cent (14.96 ± 1.11) µ moles substrate hydrolyzed/30 min/mg protein of the control value following exposure of 0.001, 0.005 and 0.010 ppm concentrations of fenoxycarb respectively.

Chapter 4

Discussion

The findings of the present research work, for the first time, reveals several essential information concerning some of the so far unexplored fenoxycarb induced changes in the life cycle stages, growth duration, longevity of adults and reproductive behaviour *i.e.* fecundity, fertility and sterility as well as in the biochemical constituents like total protein, total free amino acids, DNA, RNA, glycogen and activity of acid and alkaline phosphatases in the testes, ovaries and eggs of the rice moth, *Corcyra cephalonica* (Tables 1-24 and Figures 1-38). The findings are discussed here in the light of influence of a juvenile hormone analogue *i.e.* fenoxycarb on the growth, development and metamorphosis as well as basic extrinsic and intrinsic cellular mechanisms such as transport, synthesis, degradation and storage in relation to the aforesaid biochemical constituents to come to some such conclusions which may in future help in devising ways and means for the effective control of this stored cereal pest.

In all the instars (I, II, III and IV) the toxicity of fenoxycarb increases with increase in its concentration on the developmental stages *i.e.* larvae, pupae as well as adults. In case of first instar larval treatment, in the present investigation, the larvicidal activity of fenoxycarb increases gradually up to 0.05 ppm concentration of fenoxycarb in the rearing medium but beyond that it shoots up to 66.00 ± 2.48 per cent at 0.10, 92.67 ± 1.23 per cent at 0.50 ppm and 100 per cent at 1.00 ppm concentrations of fenoxycarb (Table 1, Figure 1). This observation indicates that at the concentrations of 0.50 and 1.00 ppm, fenoxycarb behaves as insecticidal agent. Thind and Edwards (1986) reported that fenoxycarb at 1 and 5 ppm concentrations caused 100 per cent larval mortality of both insecticide susceptible and resistant strains of *T. castaneum.* The higher larvicidal activity at higher concentrations of fenoxycarb, in the present investigation, can be compared with the larvicidal activity of fenoxycarb to the last instar larvae of *Ephestia kuehniella*, where topical application of fenoxycarb at 0.1, 1 and 10 μg/ml caused extreme increase in its larval mortality (Moreno *et al.*, 1992), and also can be compared with the larvicidal activity of methoprene to the fourth instar larvae of *C. cephalonica*, where topical application of methoprene at doses of 1.0 and 0.5µ/larva caused 100 per cent mortality by 42 and 90 days respectively, after treatment. Such larval treatment resulted in the formation of supernumerary larva which died within 7-9 days (Ambika and Abraham, 1982). In addition, 10 and 100 µg of methoprene caused 11.00 and 16.67 per cent larval mortality respectively in the last larval instars of *C. cephalonica* (Deb and Chakravorty, 1985), and their larvae attained extraordinary giant size without undergoing extra moult. Similarly, Kostyukovsky *et al.* (2000) reported that 0.1, 0.5, 1, and 2 ppm of pyriproxyfen (a fenoxycarb derivative) caused 100 per cent larval mortality in F_1 generation of insecticide susceptible and actellic resistant strain of *T. castaneum* when parental adults were allowed to lay eggs in treated food medium. They also reported that methoprene at concentrations of 1 and 3 ppm caused 100 per cent larval mortality in case of *Tribolium castaneum*, and the only difference in between these two concentrations was that at 1 ppm concentration of methoprene the larval death occurred in 117 ± 5.4 days while at the concentration of 3 ppm methoprene the larvae died within 139 ± 4.3 days. Methoprene at a concentration of 10 ppm has been found to cause severe effect against the larvae of *T. confusum* (El-Sayed, 1984) and hydroprene exposure at a concentration of 5 ppm has also been reported to cause 53.8 per cent mortality to the full grown larvae of *T. confusum* (El-Sayed, 1988). El-barky *et al.* (2012) reported that pyriproxyfen caused a significantly dose-dependent enhancement in the larval mortality of *T. castaneum*. But, contrary to methoprene, hydroprene exposed *C. cephalonica* larvae at a dose level of 100 µg did not reveal larval mortality and formed 32.78 per cent larval adultoids while a dose level of 10 µg of the same compound also did not reveal larval mortality and produced 36.36 per cent larval adultoids in *C. cephalonica* (Deb and Chakravorty, 1985). Mortal effect of 5 ppm methoprene treated wheat flour was recorded against early and late instars of three stored product insects by Mian and Mulla (1982b). Methoprene during early instars at a concentration of 5 ppm caused 17.5, 60 and 65 per cent larval mortality in case of *O. surinamensis*, *T. castaneum* and *R. dominica* respectively (Mian and Mulla, 1982b) while this compound when exposed during late instars (second instar) caused 97.5, 100 and 83 per cent larval mortality in case of *O. surinamensis*, *T. castaneum* and *R. dominica* respectively (Mian and Mulla, 1982a). When penultimate instar larvae of common citrus swallow tail butterfly, *Papilio demoleus* were treated with 7.50, 15.30 and 60 µg/µl of diofenolon, a novel JHA, its topical application caused 100 per cent mortality in larval stage (Singh and Kumar, 2011).

In holometabolous insects, the developmental switch between juvenile and adult forms depends on juvenile hormone (JH), a sesquiterpenoid produced by the corpora allata gland (Gilbert *et al.*, 2000). The presence of JH in pre-final larval instars ensures that the next molt, promoted by ecdysteroids, produces another, only a larger larva (Nijhout, 1994). In the present investigation, larval mortality may be the result of competing of JHA with natural JH for binding to its receptors or to the JH carrier proteins, injuring the corpora allata cells, or interfering with JH biosynthesis (Leighton *et al.*, 1981). Disturbed biometrical proportions of different parts of the body of affected larvae may also contribute to the ecdysial failure (Slama *et al.*, 1974). Larval mortality however, may be also attributed to combination of secondary factors, which are not directly related to the hormonal ability of JHAs (Sehnal, 1983) but may also be related to the suffocation, bleeding and dessication due to imperfect excuviation, starvation due to morphogenetic effects, failure of vital homeostatic processes and metabolic impairment. Loss of normal activity due to JHA treatment can also be correlated with muscle paralysis (Mulder and Gijswijt, 1973), which could be the direct cause of larval mortality in any instar (Retnakaran and Wright, 1987 and Soshkin, 1991).

In the present investigation, it was observed that due to increased larval mortality pupation decreased with increase in the fenoxycarb concentration. Upto 0.01 ppm concentration, pupation was not severely affected by the action of fenoxycarb but beyond this concentration pupation was severely affected in dose-dependent manner. The concentrations 0.10 and 0.50 ppm fenoxycarb are considered to be much toxic, as the rate of pupation was very low *i.e.* 34.00 ± 2.48 and 7.33 ± 1.23 per cent respectively, in comparison to rest of the concentrations (Table 1, Figure 1). Similar results have also been reported in case of *C. cephalonica* following exposure of methoprene (Deb and Chakravorty, 1985) and Ro 20-3600 (Bhargava and Devraj Urs, 1992) and in *T. castaneum* treated with methoprene (Kostyukovsky *et al.*, 2000). Bhargava and Devraj Urs (1992) reported that exposure of Ro 20-3600, a juvenile hormone analogue, decreased pupation of *C. cephalonica* in dose-dependent manner when fourth or fifth instar larvae were treated. Decreased pupation with increase in concentration of fenoxycarb was also achieved by Moreno *et al.* (1992) in *E. kuehniella* and Liu and Chen (2001) in *Chrysoperla rufilabris*. Methoprene at a concentration of 0.1 ppm caused reduction in pupation of *T. castaneum* larvae (Kostyukovsky *et al.*, 2000) and these pupae failed to emerge, while the same JHA at a conentration of 10 ppm or above inhibited pupation of the larvae of flour beetles, but remained ineffective against the grain beetles (Loschiavo, 1975). In the present investigation, 0.50 ppm concentration of fenoxycarb exposed to first instar *C. cephalonica* larvae resulted in 7.33 ± 1.23 per cent pupation but these pupae did not emerge and get perished. Similar finding has also been reported by Smet *et al.* (1989) who observed that larvae of the confused flour beetle, *T. confusum*, exposed to methoprene during larval stage died during the pupal stage. A number of JHAs have also been reported to decrease pupation in both *P. interpunctella* and *C. cephalonica* (Fajardo and Morallo-Rejesus, 1980).

Pupation is normally reduced in JHA treated larvae either due to larval death during metamorphosis or inhibition of pupation due to prolonged juvenilization, when fed, injected or in contact with JHAs and/or IGRs (Mondal and Parween, 2000). Giant larvae produced by JHAs in *T. castaneum* also failed to pupate (Ishaaya and Yablonski, 1976). Pupation is found to be negatively correlated with the concentration of IGRs and positively correlated with the age of the treated larvae (Kramer *et al.*, 1985; Loschiavo, 1975; Hoppe, 1976; Mian and Mulla, 1982a; Saxena and Kumar, 1982, 1991; Eisa *et al.*, 1984; Mondal and Port, 1995 and Parween, 1996). Most JHAs at a concentration of 10 ppm or above have been reported to inhibit pupal eclosion in both beetles and moths of stored grains (Carter, 1975; Fajardo and Morallo-Rejesus, 1980 and Mian and Mulla, 1982a).

Our finding regarding pupation in *C. cephalonica* following exposure of fenoxrcarb is in accordance with the results of Loschiavo (1975) in case of *T. confusum* exposed to methoprene; Ishaaya and Yablonski (1976) in case of *T. castaneum* treated with different JHAs; Deb and Chakravorty (1985) in case of *C. cephalonica* and Smet *et al.* (1989) in case of *T. confusum* exposed to methoprene; Bhargav and Devraj Urs (1992) in case of *C. cephalonica* treated with JHA (RO 20-3600) and Kostyukovsky *et al.* (2000) in case of *T. castaneum* exposed to methoprene.

Dhadialla *et al.* (2005) reported that in Lepidoptera the low concentration of juvenile hormone coupled with 20-hydroxyecdysone titers promotes larva to pupal moults. Due to excess of JHA in insect body only a small percentage of larvae were able to metabolize this unusual high concentration of JH in its body and got success to reach at pupal stage in dose-dependent manner. A reduction in pupation may also be the result of ecdysial failure caused by the inhibition of secretion or activity of moulting hormone. Loss of normal activity can also be correlated with muscle paralysis (Mulder and Gijswijt, 1973), which could be the direct cause of larval mortality in any instar that leads to poor pupation (Retnakaran and Wright, 1987 and Soshkin, 1991).

The pupal mortality, in the present investigation, has also been recorded to increase with the increase in the concentration of fenoxycarb. In case of first instar larvae, lower concentrations of fenoxycarb *i.e.* 0.001 and 0.005 ppm caused gradual enhancement in pupal mortality while its 0.01 ppm concentration caused moderate toxicity that resulted 21.67 ± 2.82 per cent pupal mortality, but higher concentrations *i.e.* 0.10 and 0.50 ppm fenoxycarb caused severe toxicity resulting into 43.47 ± 4.13 and 100 per cent pupal mortality respectively (Table 1, Fig.2). Increased pupal mortality, with increase in concentration of fenoxycarb was also achieved by Moreno *et al.* (1992) in *E. kuehniella* and Liu and Chen (2001) in *C. rufilabris.* This result is supported by the studies which showed 95 per cent pupal mortality in case of *T. castaneum* when treated with 3 ppm concentration of methoprene (Kostyukovsky *et al.,* 2000). The result of Parthasarathy and Palli (2009) indicated that 1 ppm concentration of methoprene blocked larval to pupal metamorphosis in 85 per cent of larvae treated during the penultimate larval stage and more than 95 per cent of larvae treated during the final instar larval stage in *T. Castaneum.* Larval treatment with 10 ppm concentration of hydroprene caused considerable pupal mortality in *T. confusum; O. mercator* and *O. surinamensis* (Loschiavo, 1975). Loschiavo (1976) again observed six species of stored-product insects that were reared on diets treated with methoprene and hydroprene at concentrations of 1, 5, 10 and 20 ppm. Both compounds at 20 ppm prevented emergence of pupae of *T. castaneum* and substantially reduced it in *T. confusum.* Pupal mortality has been reported to increase with increasing concentrations of pyriproxyfen in case of *T. castaneum* (El-barky *et al.,* 2012). JHA like MV-678 at 5 ppm concentration produced lethal effects in the pupal stage of *O. surinamensis* and *R. dominica* (Mian and mulla, 1982a). IGRs often result in pupal mortality either by pupal treatment, as reported by Soltani *et al.* (1993) or by larval treatment. Exposure of C_{18} juvenile hormone has also been reported to increase pupal mortality in addition to the formation of pupal-adult intermediates in *T. molitor* larvae (Edwards, 1976). Pupal mortality is dose-dependent (Parween, 1996) and decreased with increased pupal age (Soltani *et al.*, 1993 and Edwards, 1976).

Fenoxycarb exposure to the larvae of *C. cephalonica*, in the present investigation, caused a significantly dose-dependent reduction in adult emergence in case of all the instars. The reduction in adult emergence is due to larval and pupal mortalities influenced by the fenoxycarb action. The adult emergence was recorded as 82.00 ± 0.89 per cent at 0.001 ppm concentration of fenoxycarb that decreased upto 19.33 ± 2.17 per cent at 0.10 ppm concentration of fenoxycarb. Due to 100 per cent pupal mortality at 0.50 ppm and 100 per cent larval mortality at 1.00 ppm concentration of fenoxycarb, there was no adult emergence at both these concentrations of fenoxycarb. Similar to this result, Thind and Edwards (1986) have also reported that fenoxycarb even at its lower concentrations *i.e.* 0.001, 0.01and 0.1 ppm caused considerable reduction in adult emergence of insecticide susceptible and resistant strains of *T. castaneum, C. ferrugineus, O. surinamensis* and *R. dominica*. They further reported that 1 and 5 ppm concentrations of fenoxycarb caused 100 per cent reduction of adult emergence of *T. castaneum, C. ferrugineus* and *O. surinamensis* whereas 93 and 100 per cent reduction of adult emergence of *R. dominica* was observed at 1 and 5 ppm concentrations of fenoxycarb respectively.

Williams (1956) was the first to report that JH application to young lepidopteran pupae blocked adult differentiation and the activity of IGRs against the developmental stages certainly resulted in the reduced adult emergence. Other studies with JHA have also shown similar effect of reduced adult emergence, for example when larvae of *T. castaneum* were treated with 0.001, 0.0165 and 0.033 ppm concentrations of methoprene, adult emergence was found to be reduced to 62.8, 27.9 and 4.6 per cent respectively (Wijayaratne *et al.*, 2012b). Methoprene at the concentration of 0.1 ppm completely inhibited adult emergence in S-strains of *T. castaneum* but this concentration was found to be less effective in case of R-strains of *T. castaneum*, while the concentration of 3 ppm was most effective that caused 100 per cent reduction in adult emergence of both strains (Kostyukovsky *et al.*, 2000). Similarly, aerosol application of methoprene alone and in combination with esfenvalerate (Jenson *et al.*, 2010a) and synergized with pyrethrins (Jenson *et al.*, 2010b) resulted in reduced adult emergence in *P. interpunctella*, and it was noticed that methoprene alone was highly effective in reducing adult emergence after larval exposure. Arthur (2010) also described the aerosol application of methoprene in combination with pyrethrins that resulted in reduction of adult emergence in case of *T. confusum* and *T. castaneum*. It was observed that *T. castaneum* larvae when exposed to methoprene-contaminated flour resulted in a significant lower number of surviving adults compared to controls (Tucker *et al.*, 2014a). It was further reported that when late instar larvae of *T. castaneum* and *T. confusum* were exposed to methoprene the number of adults emerged were found to be reduced in comparison to the control (Tucker *et al.*, 2014b). Hydroprene at a concentration of 5 ppm reduced adult eclosion in *T. confusum* and *T. granarium* and the effect was found to be greater in *T. granarium* than in *T. confusum* (El-Sayed, 1988). Even its low concentration has also been reported to be a high percentage inhibitor of adult emergence in *C. cephalonica* (Vijayalakshmi and Ramraj, 1991). Abo-El-Gher (1992) evaluated the effect of five JHAs *i.e.* CGA-29170, CGA-45128, MV-678, RO-20458 and fenoxycarb against the southern cowpea weevil, *C. maculatus* (F.) and observed that the adult emergence in all the treatments were significantly suppressed. Similar to our results, which revealed 100 per cent reduction in adult emergence at 24 ppm concentration of methoprene in case of *C. cephalonica* fenoxycarb also at application rate of 20 ppm on *S. cerealella* caused 100 per cent reduction in adult emergence (Eisa, 1992). In addition, increased concentrations of pyriproxyfen also resulted in decreased percentage of adult emergence in case of *P. interpunctella* (Ghasemi *et al.*, 2010); *T. castaneum* (El-barky *et al.*, 2012); and *E. integriceps* (Mojaver and Bandani, 2010). Longer exposure of eggs, larvae or pupae of *E. cautella* to methoprene, significantly reduced adult emergence (Shaaya and Pisarev, 1986). Wijayaratne and Field (2010) reported no adult emergence from the larvae of *T. castaneum*, exposed to methoprene treated varnished wood at 24 week after treatment at the temperatures of 20, 30 or 35°C. Methoprene exposure effectively inhibited adult emergence in *S. cerealella* (Stockel and Edwards, 1981) and fenoxycarb also revealed similar effects in lesser mealworm, *Alphitobius diaperinus* and corn borer, *Ostrinia nubilalis* (Edwards and Abraham, 1985), and red flour beetle, *T. castaneum* (Vinuela *et al.*, 1990). Both these JHA compounds were found to be more effective on the emergence of *T. castaneum* adults in stored peanuts than diflubenzuron (Daglish *et al.*, 1992).

Our finding pertaining to adult emergence of *C. cephalonica* following exposure of fenoxycarb is in accordance with the results of the above workers as observed in case of *C. cephalonica* exposed to hydroprene (Vijayalakshmi and Ramraj, 1991); *T. castaneum, C. ferrugineus, O. surinanamensis* and *R. dominica* treated with fenoxycarb (Thind and Edwards, 1986); *S. cerealella* exposed to fenoxycarb (Eisa, 1992); *C. maculatus* treated with CGA-29170, CGA-45128, MV-678, RO-20458 and fenoxycarb (Abo-El-Gher 1992); *T. castaneum* exposed to methoprene (Kostyukovsky *et al.*, 2000); and *P. interpunctella* (Ghasemi *et al.*, 2010), *T. castaneum* (EI-barky *et al.*, 2012) and *E. integriceps* (Mojaver and Bandani, 2010) treated with pyriproxyfen and *T. castaneum* and *T. confusum* exposed to methoprene (Tucker *et al.*, 2014a, 2014b).

At the beginning of the pupal stage of holometabolous insects, there is an additional JH-sensitive period for pupal versus adult determination that JH must be absent in epidermal cell obligated to adult development (Nijhut, 1994). Application of JHA compound to insects causes abnormal morphogenesis of the integument. In addition, JHAs do not interfere with the function and growth of insect epidermal cells, but prevent their imaginal differentiation (Sehnal, 1983). Thus, it may be concluded that in the present investgation JHA (fenoxycarb) does not interfere with the function and growth of insect epidermal cells, but prevent their imaginal differentiation (Sehnal, 1983) that cause abnormal morphogenesis of the integument resulting in the formation of deformed pupae and adults, hence reduced percentage of adult emergence.

It was also observed that even at the same concentration of fenoxycarb the larval and pupal mortality decreases, with the increase in the age of larva at the time of treatment (Table 1-4, Figures 1-8). In the light of present findings it may be concluded that the toxicity of fenoxycarb is directly related to the age related duration of treatment and therefore early instars were more susceptible in comparison to the late instars.

The principal hormones involved in the life processes of insects include neurohormones (neuropeptides), ecdysteroids (molting hormones) and the sesquiterpenoid juvenile hormones (JHs). In general, ecdysone (hydroxylated steroid) is involved in molting while juvenile hormone (sesquiterpene) is involved in maintaining the insects in current form (Status quo) (Hoffmann and Lorenz, 1997; Gade *et al.*, 1997 and Goodman and Granger, 2005). In insects with complete metamorphosis at the time of the final molt, JH is absent and the adult emerges. Therefore, persistence of JH or juvenile hormone analogue (JHA) during that time, depending upon the dose and time of application, gives rise to the various abnormalities in larva, pupa and adults *i.e.* formation of giant larvae, supernumerary larvae, larval-pupal intermediates, malformed pupae and abnormal adults (Mojaver and Bandani, 2010).

Application of Juvenile hormone or its mimics often results in prolongation of larval period either by repetition of larval moult or by formation of supernumerary or giant larvae. Fenoxycarb, being carbamate caused a very different influence. Its higher concentrations of *i.e.* 0.01, 0.05, 0.10, 0.50 and 1 ppm produced giant larvae, supernumerary larvae, larval-pupal intermediates and abnormal pupae in case of first instar. Similar observation has also been reported by Edwards (1976) in case of *Tribolium castaneum* larvae exposed to juvenoids for a longer period which frequently led to the production of giant larvae. This finding was also supported by Dyte (1972) who observed that methoprene exposure for a tenure of 12 days produced giant larvae in *T. castaneum* that were heavier and darker than untreated larvae of the same age group. The effect was found to be similar in both susceptible (FSS II) and resistant (Kano and CTC 12) strains of the beetle, *T. castaneum* (Dyte, 1972). According to Hoppe (1976) exposure of Juvenoids such as Ro 10-3108 and Ro 20-3600 produced giant larvae in *T. castaneum* but not in *E. cautella*. These giant larvae were characterized by large head capsules and they completed development into morphologically normal appearing adults when transferred to untreated food.

Fenoxycarb showed the potential for prolonging the larval stages and formation of supernumerary larvae or larval-pupal intermediates. Deb and Chakravorty (1985) also found extra larval instars, larval-pupal intermediates and adultoid larvae of *C. cephalonica* by implantation of additional corpora allata into the last instar larvae of this insect. Applications of fenoxycarb in the fifth instar produced different effects, depending upon the dose and the timing of application, which resulted in production of supernumerary or permanent larvae, or of larval-pupal intermediates (Gadenne *et al.*, 1990). Such findings were also achieved by Moreno *et al.* (1992) when 0.1, 1 and 10 ppm of fenoxycarb was applied topically to *E. kuehniella*. Induction of extra larval moult due to topical treatment of fenoxycarb was also reported by Kamimura and Kiuchi (2002) on third and fourth instar larvae of silkworm. They mentioned that the female larvae were more likely to molt extra than males. Prolongation of larval period due to treatment of JHA to penultimate instar larvae of *P. demoleus* has also been reported by Singh and Kumar (2011). Such findings are in accordance with the effect of 6, 7, epoxy-3 ehtyl-1 (p- ethyl phenoxy)-7 methynonane treatment on one day old larvae of *E. fabia* (Srivastava and Srivastava, 1981).

Application of high concentration (1000 ppm) of methoprene increased larval life of *T. castaneum* by 43 ± 1 days that resulted into formation of superlarvae (Hoppe, 1981). Methoprene and hydroprene at the concentration of 5 ppm prolonged the larval period of *Tribolium* species up to 120 and 150 days respectively, whereas, untreated larvae pupated in 25 days (Loschiavo, 1976). He concluded that many larvae failed to pupate in treated food and continued to molt, and survived upto 120 days or more, and after emergence were larger in size than the adults emerged from normal larvae. The juvenile hormone analogues ZR-512 and ZR-619 at concentrations of 10-1000 ppm induced prolongation of the larval feeding period upto tenfold, increasied larval weight upto double than that of untreated larvae in case of *T. castaneum* (Ishaaya and Yablonski, 1976). Low doses of hydroprene (0.1 and 0.75 μg) have been found to increase the total developmental period by 18 days in *C. maculatus* larvae (Rup and Chopra, 1984). Pyriproxyfen increased larval duration with increase in its concentration in case of *T. castaneum* when exposed at fourth instar larval stage (El-barky *et al.*, 2012). In addition to methoprene, MV-678 (a JHA)) has been found to be effective against larvae of different stored cereal insect-pests, their larval body wall became loose and shrinked, stoped feeding and normal movement, gradually turned black and then ultimately died (Mian and Mulla, 1982a). In the present investigation, fenoxycarb treated larvae after a variable period shrinked their body wall, became loose, stoped their feeding and movement, gradually turned black and ultimately died. This finding is in accordance with the results of Mian and Mulla (1982a) as observed in case of coleopterans like *T. castaneum, R. domonica, S. oryzae* and *O. surinamensis* following exposure of JHAs. The induced prolongation of the larval stage after juvenile hormone treatment was followed by a pronounced enhancement of cuticle phenoloxidase activity, indicating an alteration of the larval biochemical processes (Ishaaya and Yablonski, 1976).

Since different tissues have a different threshold of sensitivity to the JH and not all the tissues reach the pupal characteristics at the same time (Nijhout, 1994), treatment with JHA can cause the development of larva-pupa or larva-adult intermediate individuals. Deb and Chakravorty (1985) has described the larval-pupal intermediates in *C. cephalonica* whose head and thorax were almost totally larval and abdomen was pupal. Singh and Johnson (2013) reported the production of such intermediates due to fenoxycarb exposure in *Alphitobius diaperinus*. Formation of larval-pupal intermediates by the application of JHA have also been reported by Srivastava and Srivastava (1981) in *E. fabia* and by Singh and Kumar (2011) in *P. demoleus*. In insects at an appropriate stage, a natural drop of JH secretion permits metamorphosis. At this critical time if excess of JHA is provided to insect it may disrupt normal developmental pathway causing repetition of larval or pupal instars respectively (Fukuda, 1944; Wigglesworth, 1954; Williams, 1956). JHAs may disturb larval growth by affecting larval weight and length, increasing larval period and morphologically abnormal mouthparts causing starvation leading to death of the larva at different instars.

The hormonal regulation of metamorphosis in Lepidoptera is described for many species (Tanaka *et al.*, 1989; Jones *et al.*, 1990 and Niimi and Sakurai, 1997). In all these Lepidoptera, the JH level is low during the last larval instar and declines prior to cocoon spinning also. In fact JH is present during larval life to allow growth of the larva and progression from one larval stage to the next until the larva attains a proper size for metamorphosis. During this period larvae are highly sensitive to JHA (Sehnal, 1983). In fact the role of JH in premetamorphic insects is to modulate the cell's response to the molting hormone, 20- hydroxy-ecdysone. Exposure to JH alone apparently has little or no effect on the cells of larval insects. At molecular level, the drastic decrease of JH initiates the metamorphosis. This initiation means a switch on the expression of larval characters to pupal characters. Genes involved in the expression of specific larval proteins will be turned off, genes involved in the expression of specific pupal proteins will be turned on. If JH is injected at this critical moment, the switch does not occur (Truman and Riddiford, 1974 and Riddiford and Truman, 1978) that show the primer role of JH in the maintainance of larval characters. In molecular level, JH regulates the turning on and the turning off of the broad transcription factor (BR). BR is the key factor to the switch from specific larval characters expression to specific pupal character expression (Zhou and Riddiford, 2002). Thus, it may be concluded that in the present investigation, fenoxycarb interferes with the broad transcription (BR) factor, which delays/or inhibits the expression of specific larval protein and consequently failure of expression of specific pupal proteins resulting into inhibition of transition/expression of larval character to pupal character. It appears that the abnormal proportion of JH in insect's body enhances larval and pupal mortality and hence there is reduced percentage of adult emergence of *C. cephalonoca*.

In this study, we recorded larvae that incompletely switched to pupae or adults, but died as pupae, whose abdomen had larval characteristics, or prepupae or pupae, with the pigmentation typical of adults. Similar responses with fenoxycarb were also observed in other insects, for example *B. mori* and *Chrysoperla carnea* (Cappellozza *et al.*, 1995 and Bortolotti *et al.*, 2005).

In the present investigation, increasing the concentration of fenoxycarb resulted in low emergence of normal adults and some pupoids and adultoids were also observed which were more pronounced at its higher concentrations *i.e.* 0.01, 0.05 and 0.10 ppm fenoxycarb in first instar treated larvae and at 0.05 and 0.10 ppm fenoxycarb concentration in second, third and fourth instar treated larvae but the number of adultoids at these concentrations were more in first instar treated larvae that decreased dose-dependently in second, third and fourth instar treated larvae, possibly due to difference in age at the time of treatment and decreased tenure of fenoxycarb exposure from first instar treated larvae to fourth instar treated larvae. Abnormilities includes twisted wings and twisted legs and abnormal abdomen. It also deserves mention that at 1.00 ppm concentration of fenoxycarb, in first instar exposed larvae, pupoids and adultoids did not form but extra moults occur resulting into the formation of giant larvae and supernumerary larvae which after certain life-span stoped feeding and movement, their body shrinked, body wall became loose, turned black and died. Deformities produced in different stored-product insects due to JHA treatments were reported earlier by Thomas and Bhatnagar-Thomas (1968) and Williams and Amos (1974). These effects are ultimately expressed either by the arrest of larval development or by deranged normal adult development.

Abnormalities in pupae and adults due to treatment with different JHAs have also been reported by various workers like effect of methoprene on *S. oryzae* and *S. granarius* (Baker and Lum, 1976); methoprene and hydroprene on *Tribolium* spp. and *Oryzaephilus* spp. (Loschiavo, 1976); methoprene on *T. castaneum* (Hoppe, 1981); hydroprene and R-20458 on *S. oryzae* (Gupta and Mkhize, 1983); hydroprene on *C. maculatus* (F.) (Rup and Chopra, 1984); methoprene on C. cephalonica (Deb and Chakravorty, 1985); methoprene and hydroprene on *T. confusum* and *T. granarium* (El-Sayed, 1984, 1988); MV-678 and R-20458 on *Tribolium* spp. (Mkhize, 1988); RO 20-3600 (fenoxycarb) on *C. cephalonica* (Bhargava and Devraj Urs, 1992); pyriproxyfen on *P. interpunctella* (Ghasemi *et al.*, 2010); and *T. castaneum* (El-barky *et al.*, 2012). These investigations suggest that application of JHAs caused morphological abnormalities in the developmental stages *i.e.* deformed pupae and adults. In addition, these JHAs applied during late larval instars, resulted in the formation of larval-pupal and pupal adult intermediates in few of cases (Bhargava and Devraj Urs, 1992).

Fajardo and Morallo-Rejesus (1980) observed effects of JHAs and BPUs on *C. cephalonica* and *P. interpunctella* which resulted in various types of abnormalities in adults and pupae. The pupal abnormalities included individuals with: larval characteristics to the anterior portion of the body but pupal characteristics to the abdomen; head and thoracic legs were larval and rest of the body was pupal; pupal head with larval thoracic legs; and constriction on the first one or two segments of the abdomen while adult deformities ranged from one or more wings being crumpled to hind wings flipped over forewings and short and vestigial wings (Fajardo and Morallo-Rejesus, 1980).

It is thought that JHAs interact with DNA molecule and as a result, either the expression of the adult genes is inhibited (Williams, 1961), the larval genes are activated (Wigglesworth, 1961), or the replication of DNA in the larval part of the body is induced (Novak, 1967), which further results the retention of larval gene traps in the pupal stage or retention of pupal gene traps in the adults (Retnakaran *et al.*, 1985), which may be possible reason of observed abnormalities in pupae and adults form in the present investigation. JHAs also influence the endocrine physiology of the insect which may, in part, cause abnormal morphogenesis and is primarily seen during the larva-pupal transformation (Retnakaran *et al.*, 1985). The degree of morphogenetic effects produced by JHAs differs with the mode of application, dose administered, species and age of the treated insects (Mondal and Parween, 2000).

JHA as a whole distrupt insect metamorphosis, hence, the developmental time for pupation and adult eclosion are considerably increased. Due to prolongation of larval period there was delayed pupation or adult emergence. Hence, overall duration of growth period was found to increase with increase in concentration of fenoxycarb in this experiment (Table 6-9; Figures 9, 11, 13, and 15). A short exposure to IGRs at definite instars of insect larvae would not affect the developmental time but a long exposure usually showed profound effects (Mian and Mulla, 1982b). In the present investigation, application of fenoxycarb to the first instar larval stage of *C. cephalonica* caused significant influence on its growth duration (the time elapsing between egg laying to adult emergence). Enhancement in the growth duration was observed when first instar larvae were treated with higher concentration of fenoxycarb (Table 6, Figure 9). The growth duration of rice moth in control was observed as 37.83 ± 0.70 days for males and 45.83 ± 0.87 days for females while this duration of first instar larvae treated with 0.10 ppm was 85.17 ± 1.81 days for males and 86.50 ± 1.98 days for females. In all instars the larval duration, at 0.5 ppm fenoxycarb concentration, was much prolonged and maximum larval duration was achieved at 1.00 ppm of fenoxycarb which was 105.67 ± 3.51days. At both these concentrations adults did not emerged. Similarly, methoprene at a dose level of 10 and 100 µg also influenced the time required for emergence of *C. cephalonica* when treated at 0-24 h old larval stage (Deb and Chakravorty, 1985) but methoprene at a concentration of 0.5 ppm was found to be poorly effective in case of resistant strains of *T. castaneum* while its 0.1 ppm concentration was not effective (Kostyukovsky *et al.*, 2000). Pyriproxyfen, a fenoxycarb derivative juvenile hormone analogue, with increased concentrations also increased growth duration effectively in case of the *P. interpunctella* even at poor concentrations (Ghasemi *ET AL.*, 2010) but the same compound was found to be poorly effective in case of *E. integriceps* (Mojaver and Bandani, 2010). Prolongation in the developmental period of *C. maculatus* has also been reported following exposure of hydroprene (Rup and Chopra, 1984). From these reportings, it is evident that toxicity of IGRs to developmental stages of insects is species specific. Since, in the present investigation, the duration of fenoxycarb exposure to the first instar larvae was more in comparison to that of second, third and fourth instar larvae, hence, the growth duration was found to be maximum in case of first instar treated larvae. Thus, even at the same concentration of fenoxycarb the growth duration of the insect increases with the decrease in the age related duration of exposure (Table 6-9, Figures 9, 11, 13, and 15). Fenoxycarb when exposed to second and third instar larvae of *O. nubilalis* had no significant effects on the duration of these instars while the duration of the first instar treated larvae increased significantly (Gadenne *ET AL.*, 1990).

Body size regulation involves the careful integration of mechanisms that control growth rate with those that control growth duration. In insects, developmental hormones such as juvenile hormone and ecdysone regulate developmental transitions and growth duration. The conserved insulin signaling pathway regulates growth rates. There is an intimate link between the three, whereby juvenile hormone controls body size by regulating ecdysone synthesis, which in turn modifies insulin signaling. The role of juvenile hormone (JH) in regulating the timing and nature of insect moults is well-established. Evidences suggest that JH is also involved in regulating final insect size (Mirth *et al.*, 2014). The authors elucidate the developmental mechanism through which JH regulates body size in developing *Drosophila* larvae by genetically ablating the JH-producing organ, the corpora allata (CA) and found that larvae that lack CA pupariated at smaller sizes than control larvae due to a reduced larval growth rate. Neither the timing of the metamorphic moult nor the duration of larval growth was affected by the loss of JH. They further stated that the effects of JH on growth rate are dependent on the forkhead box O transcription factor (FOXO), which is negatively regulated by the insulin signaling pathway. Larvae that lacked the CA had elevated levels of FOXO activity, whereas a loss-of-function mutation of FOXO rescued the effects of CA ablation on final body size. Finally, the effect of JH on growth appears to be mediated, at least in part, via ecdysone synthesis in the prothoracic gland. These results indicate a role of JH in regulating growth rate via the ecdysone and insulin signaling pathways. In the present investigation it appears that fenoxycarb interferes with ecdysone synthesis and hence enhances the growth duration of *C. cephalonica*.

Prolongation of larval period by application of JHA in food medium was reported by Kostyukovsky *et al.* (2000), Williams and Amos (1974), Liu and Chen (2001) and Ghasemi *et al.* (2010) at different concentrations. Methoprene at dose levels of 10 and 100 µg also influenced the time required for emergence of *C. cephalonica* when treated at 0-24 h old larval stage (Deb and Chakravorty, 1984). It is noteworthy that in the present investigation growth duration of female was slightly longer than male in control as well as in treated group. In all the instars maximum larval duration was achieved at 1.00 ppm concentration of fenoxycarb which was 105.67 ± 3.51 and 103.67 ± 2.95 days in case of first and second instar treated larvae respectively (Table 6 and 7). The intensity of increase in larval duration was decreased with older instar larval treatments as at 1.00 ppm concentration of fenoxycarb larval duration was 102 ± 2.66 and 97.17 ± 2.96 days for third and fourth instars respectively (Table 8 and 9). These larvae remained as larvae and after variable periods they stoped feeding and movement, turned, black and eventually died.

In the present work, the longevity of *C. cephalonica* adults emerged from fenoxycarb treated larvae was significantly reduced (Table 6-9, Figures 10, 12, 14 and 16). In the control, adult longevity was 13.50 ± 0.42 days for males and 8.50 ± 0.43 days for females whereas the longevity of adults emerged from first instar larvae treated with 0.10 ppm fenoxycarb was 5.00 ± 0.86 days for males and 4.00 ± 0.97 days for females. Similar findings were reported following the exposure of pyriproxyfen on *P. interpunctella* exposed to 0.3 ppm of pyriproxyfen (Ghasemi ET AL., 2010) and on *E. integriceps* (Mojaver and Bandani, 2010). When 6-day-old *T. molitor* pupae were treated with 0.1 µg JHA, adult life span was reduced to 4–6 days against the normal *i.e.* 14–16 days (Metwally and Landa, 1972). Methoprene and hydroprene also at the concentrations of 0.25, 0.5, 1 and 2 ppm reduced longevity in case of *C. cephalonica, C. maculatus and T. castaneum* (Murali Baskaran and Janarthanan, 1991). However, methoprene and pyriproxyfen act in a different way *i.e.* they have no effect on the life-span of *R. dominica* and *S. oryzae,* but have a profound effect on *T. castaneum* (Kostyukovsky *et al.,* 2000). IGRs are not toxic to the adults themselves but long exposure to the early larval stages cause disruptions in the organ systems or interrupts adult ecdysis and reduces the adult life-span (Mondal and Parween, 2000). Fenoxycarb reduced adult life-span in other species than did malathion in stored rice (Cogburn, 1988). Tucker *et al.* (2014b) also determined the ability of *T. castaneum* and *T. confusum* to develop successfully when treated at late instar larval stage with methoprene. They stated that for *T. castaneum,* survival time for individuals exposed to methoprene was shorter than those exposed to control individuals, but the difference was singnificant for *T. confusum.*

Adult longevity also depends on healthy immature stages. Digestive disorders such as starvation, disturbance in metabolism, degeneration of peritrophic membrane and accumulation of faecal materials at the hind gut may be the cause of untimely adult mortality as a result of BPU exposure (Soltani, 1984 and Parween, 1997). Mean longevity of mated adults treated with 10 ppm IGR (juvenoids and BPUs) was decreased compared to untreated and unmated adults in *P. interpunctella* whereas the tested IGRs lengthened life-span of the mated adults of *C. cephalonica* (Fajardo and Morallo-Rejesus, 1980). The toxic effect of fenoxycarb might be responsible for the reduction in the longevity of adults from treated culture (Table 6-9, Figures 10, 12, 14 and 16).

The longevity of adults has also been found to be increased with the increase in the age related duration of treatment. Adults emerged from first instar larvae treated with fenoxycarb survived for shortest time in comparison to those adults emerged from second, third and fourth instars treated larvae.

The observations, presented in the Tables 10-13 and Figures 17, 19, 11 and 23 clearly indicate that the egg laying capacity of the female decreases with the increase in the concentration of fenoxycarb. The maximum inhibition of egg laying capacity was observed in the crosses between treated females and treated males whereas minimum inhibition was recorded in the crosses between normal females and treated males. It was observed that increase in the concentration of fenoxycarb caused severe effect on the reproductive behaviour of the *C. cephalonica.* Similar observations have been recorded after larval treatment with different concentrations of methoprene in *T. castaneum* and *T. confusum* (Loschiavo, 1976); *R. dominica* (Daglish and Pulvirenti, 1998; Chanbang et. al., 2008); *T. castaneum* (Wijayaratne *et al.*, 2012a) and *C. medinalis* (Sun *et al.*, 2013). Other juvenoids like pyriproxifen has also been reported to inhibit fecun-dity in *P. interpunctella* (Ghasemi *et al.*, 2010). It was also recorded, in the present investigation, that even at the same concentration of fenoxycarb, the suppression of egg laying capacity of the females decreases with the increase in the age related duration of exposure (Tables 10-13 and Figures 17, 19, 11 and 23).

JHA suppresses egg production in insects by several mechanisms. In males JHAs disrupt spermatogenesis (Dumser and Davey, 1974) and functioning of accessory glands in some insect species and the degree to which the target tissue is affected differs with the species. In females, JHAs affect the development of oviducts (Koeppe *et al.*, 1985), follicular growth (Koeppe *et al.*, 1980), oocyte maturation (Koeppe *et al.*, 1985) and functioning of accessory glands (Koeppe *et al.*, 1985 and Bodenstein and Sprague, 1959). Morphology of genitalia is also affected by external exposure of methoprene (DeVries and Brown, 1977). JHAs treatment produced severe disorders in the ovaries of *T. molitor* including cell death in the germarium, resorption of oocytes in the previtellarium and vitellarium, formation of compound egg chambers and undue proliferation of follicular cells sometimes resulting in malformation of the whole ovaries (Metwally *et al.*, 1972). These ovarian defects appear to be consequent with the changes in the course of cell proliferation, and JH compounds seemed to interfere with the cell division and differentiation (Metwally *et al.*, 1972). In addition, the absence of JH at the end of larval life is necessary for the development of the normal reproductive behaviour of the adults. Interference of JH in the reproductive behaviour could be the direct cause of decreased fecundity in treated females (Edward and Menn, 1981).

Reproduction in insects is mainly controlled by the corpus allatum hormone, which is also responsible for protein metabolism and is specifically needed for egg maturation. JHAs have been found to render treated insects either sterile or less fecund (Metwally *et al.*, 1972). IGR treated insects may develop as morphologically deformed adults which would be non-viable or at least their reproductive capacity are reduced (Williams and Amos, 1974).

Earlier findings suggest that in pre-adult females of *S. oryzae*, hydroprene and R-20458 treatment prevented successful completion of vitellogenesis, formation of the germinal vesicle and karyosphere, and also blocked retraction of the oolemma and as a result the ovarioles remained atrophied and consequently no new generation was produced (Gupta and Mkhize, 1983). Diflubenzuron reduced oviposition in *T. molitor,* as a result of inhibition of ovarian DNA synthesis (Soltani-Mazouni and Soltani, 1994) and also due to interference with vitellogenesis via biochemical processes (Oberlander *et al.*, 1975; Soltani *et al.*, 1984 and Soltani-Mazouni, 1994). Fenoxycarb at 0.1-10 ppm concentrations caused abnormal development of the male gonads in *E. kuehniella* (Moreno *et al.*, 1992) resulting into reduced fecundity of adults. In addition, methoprene and hydroprene treated diet reduced the viability of deformed females, and that of both normal and deformed males of *T. castaneum* (Amos *et al.*, 1978). Larval diet of triflumuron disrupted the testicular follicle in male *T. castaneum* of both susceptible and resistant strains (Parween, 1996). Spermatids and sperm bundles in the treated males appeared to be abnormally large, and spermatogonia were few in number. When these treated males were allowed to mate with untreated females, egg viability was reduced and the effect was found to be dose-dependent. However, the male reproductive system of *T. castaneum* was less affected than the female reproductive system. At higher doses of IGR treatment sterility may be permanent but at marginal doses the females may lay viable eggs at the later part of the oviposition period (Adan *et al.*, 1994 and Parween, 1996). The ability of JHA-treated males to affect fecundity in females, has also been reported in *Pyrrhocoris apterus* (Masner *et al.*, 1968), *Trogodema granariurn* (Metawally and Landa, 1972) and *C. fumiferana* (Hicks and Gordon, 1992).

The hatchability of eggs collected from the different crosses was also recorded to be reduced with increase in the concentration of fenoxycarb in the rearing medium. The maximum reduction of hatchability was observed in the crosses between treated males and treated females and minimum reduction was recorded in the crosses between normal females and treated males (Tables 10-13 and Figures 18, 20, 22 and 24).

Most IGRs are potential ovicides and are reported to affect hatchability, either by direct treatment of the eggs or as a delayed effect by treating the parents at any stage of their life-cycle (Mondal and Parween, 2000). Similar findings have also been reported for hydroprene and methoprene treated *S. granarius* (Nordlander, 1975), hydroprene treated *C. maculatus* (Rup and Chopra, 1984) and *R. dominica* when their eggs were exposed to methoprene (Chanbang *et al.*, 2008). Reduced hatching of eggs were also reported by Mojaver and Bandani (2010) in *E. integriceps* and by Ghasemi *et al.* (2010) in *P. interpunctella* when treated with pyriproxyfen.

During treatment, age of the insect/larva and their duration of exposure to the IGRs are also important factors in the hatchability of eggs. It was observed in the present investigation that hatchability of eggs decreases with the increase in the duration of treatment *i.e.* from fourth instar to first instar larval treatment.

It is also clear from the oviposition and hatchabilty data that even at the same concentrations of fenoxycarb, maximum inhibition in the egg laying capacity and hatchability of eggs were observed in the crosses between treated males and treated females wereas minimum inhibition of that was recorded in the crosses between normal females and treated males (Table 10-13, Figures 17-24). It shows that females are more susceptible to methoprene treatment than the males.

Slama and Williams (1966) also reported reduced hatching in JHA treated eggs which were either due to failure to close the dorsal suture connecting the embryo with the yolk sac and muscle atrophy in the appendages or due to toxic effect which kill the embryo within the eggs. Disruption of embryogenesis by JHAs has been demonstrated in several other insect species (Retnakaran *et al.*, 1985; Charmillot *et al.*, 1985 and Masner *et al.*, 1987). In addition, the effect of terpenoid and nonterpenoid JHAs in impairing the fecundity of adult females has been documented for a variety of insect species (Retnakaran *et al.*, 1985). They block embryonic development by inhibiting the differentiation of the embryos (Masner *et al.*, 1968 and Edwards and Menn, 1981), usually before or at blastokinesis (Slama and Williams, 1966; Riddiford and Williams, 1967; El-Ibrashy, 1970 and Benskins and Vinson, 1973) or affect yolk deposition in the treated eggs (Slama, 1971). These compounds have been observed to disturb the growth and maturation of the gonads in both sexes (Metwally *et al.*, 1972; Soltani-Mazouni, 1994; Soltani and Soltani-Mazouni, 1995 and Parween, 1996) resulting in production of non-viable eggs. Our findings regarding the fecundity and fertility are in conformity of Ghasemi *et al.* (2010), Wijayaratne *et al.* (2012a) and Chanbang *et al.* (2008) as reported in case of *P. interpunctella, T. castaneum* and *R. dominica* respectively.

The sterility in *C. cephalonica* is recorded in terms of inhibition of egg laying capacity and reduced egg hatchability. Tables 10-13 and Figures 17-24 clearly indicate that significant sterility can be achieved by treating larval stages of this moth with sub lethal concentrations *i.e.* 0.001, 0.005 and 0.01 ppm fenoxycarb in rearing medium and maximum sterility can be achieved by treating first instar larvae of this insect.

Sterility in insects might be due to (i) distorted gonads (ii) inhibition of vitellogenesis in the ova of treated females (iii) reduction of the number of spermatophores in treated males and (iv) extrusion of ovipositor in the adult females (Mondal and Parween, 2000). Treatment of immature stages with IGRs produced various degrees of ovarian abnormalities, and adults with such abnormality were generally sterile (Mkhize and Gupta, 1982 and Deb and Chakravorty, 1981). The permanent ovarian abnormalities led to female sterility (Metwally and Landa, 1972; Metwally *et al.*, 1972 and Deb and Chakravorty, 1981). Reproductive capacity was found to be reduced due to decreased ovariole number and reduced length of the ovary in JH-treated Colorado potato beetle, *Leptinotarsa* decemlineata (Ammar, 1983). In pre-adult females of *S. oryzae*, hydroprene and R-20458 treatment, either topically or through treated food, prevented successful completion of vitellogenesis, formation of the germinal vesicle and karyosphere and blocked retraction of the oolemma. As a result the ovarioles remained atrophied and consequently no new generation was produced (Gupta and Mkhize, 1983).Untreated females can also be sterilized through sexual contact with the treated males (Parween, 1996). Repeated mating of untreated females with treated males gave a more pronounced effect (Masner *et al.*, 1970). Both JHAs and CSIs are reported to possess sterilizing effects mainly on the females (McGregor and Kramer, 1976; Parween, 1996). Similar reasons may be accounted for reduced fecundity, fertility and enhanced sterility of *C. cephalonica* treated with methoprene as observed by a number of workers in different insect species with the application of various IGRs (Metwally and Landa, 1972; Metwally *et al.*, 1972; Deb and Chakravorty, 1981; Mkhize and Gupta, 1982; Ammar, 1983; Gupta and Mkhize, 1983 and Mondal and Parween, 2000).

When first instar larvae of *C. cephalonica* were exposed to sub lethal concentrations *i.e.* 0.001, 0.005 and 0.010 ppm fenoxycarb, certain biochemical constituents of gonads and eggs (laid within 24 h of mating) were observed to exhibit significant quantitative variations.

Proteins are among the most complex of all known chemical compounds and also the most characteristic of living organisms. They serve as an important internal environmental factor for the metabolism, especially having a close relation with fat body, metamorphic hormone, trehalose and sex hormone during development and metamorphosis (Lee *et al.*, 1981).

The total protein content of the testes and ovaries, in the moths emerged from the fenoxycarb treated cultures, was found to be reduced dose-dependently ($p < 0.001$) and this reduction was observed to be increasing with increase in the concentrations of the fenoxycarb in rearing medium (Table 14, Figure 25). Similar to the present finding, pyriproxyfen at the doses ranging from 0.02 to 0.3 ppm has been found to cause a significant reduction in the total protein content in the ovaries of *P. interpunctella* (Ghasemi *et al.*, 2010) and in the testes and ovaries of *C. stolli* exposed to juvenile hormone analogue (Saha *et al.*, 1985). Tebufenozide, an ecdysteroid agonist, also caused a significant reduction in the protein content of the testicle of *E. kuehniella* while in their ovaries, protein levels were unaffected by the exposure of this compound (Khebbeb *et al.*, 2008). In addition, chlorfluazuron, a chitin synthesis inhibitor, exposure also resulted in a singnificant decrease in the total protein content in the ovaries of common cutworm, *S. litura* (Perveen and Miyata, 2000 and Perveen, 2012) as well as in the testes of this insect (Perveen, 2012). They speculated that the decrease in ovarian protein content in chlorfluazuron treated females was resumed to have several causes, such as the lack of protein in the ovarioles or interference of chlorfluazuron with the mechanism controlling yolk deposition. They explained that the decrease in ovarian protein content suggests an interference of this IGR with vitellogenesis. High doses (1.25, 2.5, 5 and 10 µg) of pyriproxyfen inhibited the onset and accumulation of vitellogenin (a protein produced by fat body cells) in the haemolymph of young worker bees in a dose-dependent fashion. This inhibition was not a result of fat body cell degeneration or death but by impairing vitellogenin synthesis (Pinto *et al.*, 2000). Etebari *et al.* (2007) reported that the amount of total protein in the haemolymph of silkworm larva at 24 h after treatment by the juvenile hormone analogue pyriproxyfen did not show significant difference with control but at the 5th day of fifth instar, total protein was observed to be reduced considerably. Hexaflumuron and lufenuron, which are chitin synthesis inhibitors, also caused a significant reduction in the total protein level in the homongenate of third instar larva of *M. domestica* (Assar *et al.*, 2010). Total protein level in whole larva of *P. interpunctella* was similarly reduced by 20-hydroxyecdysone (Rharrabe *et al.*, 2009).

Similar findings have also been reported in other insect species by different IGRs, such as, pyriproxyfen (El-Sherif, 1995) and lufenuron (Bakr *et al.*, 2007) against *S. gregaria*; pyriproxyfen (Shaurub *et al.*, 1998) and tebufenozide (Abd El-Mageed, 2008) against *S. littoralis*; pyriproxyfen against *A. ipsilon* (El-Sheikh, 2002) and by pyriproxyfen and hexaflumuron against *P. aegyptiaca* (Assar, 2004). They reported that the decline in protein content obtained by lufenuron and hexaflumuron can be explained according to Mitilin *et al.* (1977) by the inhibition of protein synthesis as a result of inhibition of DNA synthesis as the first sign of cell death. Contrary to the present finding, FMC-23509 (CRD 9499), a juvenoid caused a significant increase in total protein level in testes and ovaries of *G. gryllotalpa* (Mandal, 1982).

The reduction in total protein level, in gonads of the present investigation is in accordance with the results obtained by Saha *et al.* (1985) on testes and ovaries of *C. stolli*; Perveen and Miyata (2000) on ovaries of *S. litura*; Khebbeb *et al.* (2008) on ovaries of *E. kuehniella* and Ghasemi *et al.* (2010) on ovaries of *P. interpunctella*.

Ahi and Rai (2012) reported that fenoxycarb treatment caused various deformities in male gonads of *Poekilocerus pictus*, such as basement membrane was distorted, the spermatogonia were pycnotic and degeneration of the primary as well as secondary spermatocytes and dissociation of spermatozoa were seen. The spermatozoa were dissociated from their bundles. During later stages the cyst cells were no longer visible, the cytoplasm was weak and clumping of chromatin threads of primary spermatocytes and dissociated sperm heads were observed. The spermatogonia were enlarged and few of these enlarged spermatogonia were binucleate, some of the secondary spermatocytes were showing division with visible metaphase stage, hypertrophied spermatids and dissociated spermatozoa were observed, (Ahi and Rai, 2012). The visible damage to gonads due to the effect of fenoxycarb appeared to be degeneration of the germ cells. In this case, the germ cells and differentiating germ cells become pycnotic. In the present investigation, it may be assumed that fenoxycarb causes various deformities in male gonads of *C. cephalonica* as reported in case of *Poekilocerus pictus* (Ahi and Rai, 2012) which may be associated with reduced protein content in testes.

It may be concluded that fenoxycarb induced decrease in the level of protein in the ovaries of *C. cephalonica* in the present investigation, may affect the oogenesis and may also create deficiency in the yolk constituents of the developing oocytes. The reduction in the nucleic acid levels in the ovaries of *C. cephalonica* after fenoxycarb treatment, in this study, may be one of the possible reasons for the decreased protein content in the ovaries. The decline in the protein content, in the present investigation, may also be due to inhibition of protein synthesis as a result of inhibition of DNA and RNA synthesis as the first sign of cell death as reported in case of boll weevil, *Anthonomus grandis* (Mitilin *et al.*, 1977).

A significantly dose-dependent ($p < 0.001$) reduction in the level of total protein in the eggs, laid within 24 h of mating by different crosses involving one or the other or both sexes emerged from treated culture was observed. The total protein level, in the eggs obtained from different crosses, was also found to be reduced with increase in the concentrations of fenoxycarb in the rearing medium. This reduction was found to be maximum in the eggs collected from the crosses between treated males and treated females and minimum in the crosses between treated males and normal females (Table 15, Figure 27). It may be supposed that fenoxycarb induced reduction in the nucleic acid levels in the ovaries (Table 17, Figures 29 and 30) and eggs (Table 18 and 19, Figures 31 and 32), and depeletion of proteins and/or inhibition of amino acid incorporation into proteins in ovaries, in the present investigation, might be the possible reasons for the decreased level of total protein levels in the eggs of rice moth, *C. cephalonica*.

The free amino acids are believed to play an important role in osmoregulation (Bishop *et al.*, 1926 and Beadle and Shaw, 1950), buffering of the blood to some extent, energy production for flight and cocoon construction (Wyatt, 1961) with the predominant function of serving as units for protein synthesis (Buck, 1953) and taking part in other metabolic activities.

The total free amino acid levels, in the testes and ovaries of moths emerged from treated culture, was found to be increased dose-dependently ($p < 0.001$) with the increasing concentrations of fenoxycarb in the rearing medium (Table 14, Figure 26). Total free amino acid levels have been reported to be increased in *M. domestica* larva following exposure of pyriproxyfen (Assar *et al.*, 2010) they further observed that buprofezin (a chitin synthesis inhibitor) and tebufenozide (an ecdysone agonist) also caused a significant enhancement in the total free amino acids in the larval body of *M. domestica.* Similar observations have also been reported in case of *M. domestica* following treatment with different IGRs like dimilin, BAY SIR-8514 and altosid *i.e.* methoprene (Bakr, 1986). He suggested that the increase of free amino acids in treated stages of *M. domestica* with dimilin, BAY SIR-8514 and altosid might be due to the destruction of protein. In addition, he also suggested that these compounds may interfere in the transcription of DNA during the process of protein synthesis which may be the reason of abnormalities. Zeenath and Nair (1994) concluded that when the sixth instar larvae of *S. mauritia* were treated with a juvenile hormone anologue hydroprene, the total free amino acids concentration increased. Bakr *et al.* (1989) mentioned that the total pool of free amino acids in the larvae and pupae of *M. domestica* was increased by treatment with dimilin and BAY SIR-8514. Our present findings are in conformity of the above workers as observed in case of *M. domestica* exposed to dimilin, BAY SIR-8514 and altosid (Bakr, 1986); *M. domestica* treated with dimilin and BAY SIR-8514 (Bakr *et al.*, 1989); *S. mauritia* exposed to hydroprene (Zeenath and Nair, 1994) and *M. domestica* exposed to pyriproxyfen

(Assar *et al.*, 2010). It may be concluded that the enhancement in the total free amino acid levels after methoprene treatment, in the present investigation, may be related with the decreased protein content due to its depletion and/or inhibition of amino acids incorporation into protein in the gonads.

A significantly dose-dependent ($p < 0.001$) enhancement in the levels of total free amino acid in the eggs, laid with in 24 h of mating, by different crosses involving one or the other or both sexes emerged from treated culture was observed (Table 16, Figure 28). The degree of enhancement was found to be maximum in the eggs obtained from the crosses between treated males and treated females and minimum in the eggs of the crosses between normal females and treated males.

The enhancement in the total free amino acid levels in the eggs of *C. cephalonica* exposed to methoprene, in the present investigation, may be related with the decreased protein content in the gonads as well as in eggs that may be due to depletion of proteins and/or inhibition of amino acid incorporation into proteins in gonads.

DNA content provides an estimate of cell number whereas RNA content can be considered as an index of the capacity of organism for protein synthesis (Brachet, 1955 and Lang *et al.*, 1965). It was observed that in the present study, fenoxycarb exposure revealed a significantly dose-dependent ($p < 0.01$) reduction in the levels of DNA and RNA contents in the testes and ovaries of adults (Table 17, Figure 29). Similar to the present investigation, application of juvenile hormone analogue has been reported to reduce DNA and RNA contents in the testes and ovaries of *C. stolli* (Saha *et al.*, 1985). Chlorfluazuron, a chitin synthesis inhibitor, has also been reported to inhibit DNA and RNA synthesis in testes and ovaries of *S. litura* (Perveen, 2012). However, the author suggested that the rate of inhibition of DNA synthesis was found to be greater than that of RNA. In addition, the total RNA decreased rapidly whereas DNA content decreased steadily and the author further concluded that DNA was more sensitive to chlorfluazuron because it has longer half-life than dimilin (Perveen, 2012). Chinzei and Tojo (1972) and Premkumar *et al.* (1991) described variations in DNA and RNA contents while studying them in *B. mori* and aquatic hemipteran bug, *Laccotrephes grisens*, respectively. Earlier findings suggest that DNA synthesis occurs only in the spermatogonia and premeiotic primary spermatocytes in, *D. melanogastor* (Das *et al.*, 1964 and Olivieri and Olivieri, 1965), *Melanoplus* (Muckenthaler, 1964), and in *A. grandis* (Chang and Riemann, 1967). Our findings are in accordance with the results as observed in case of *C. stolli* exposed to juvenile hormone analogue (Saha *et al.*, 1985) and *S. litura* treated with chlorfluazuron (Perveen, 2012).

The general growth of insects is controlled by moulting hormone and juvenile hormone whose metabolism and degradation are checked by enzymes. If degradation continues then further action by that particular hormone will stop and ultimately growth of the insect is retarded. Later it was suggested by Socha and Sehnal (1972), moulting hormone activated the synthesis of RNA and JH simultaneously induced the duplication of DNA. It means that growth hormones affect nucleic acid production and their quantity. The reduction in DNA content in testes is directly proportional to the number of sperm present in the testes. A number of researchers worked on the aspect that IGR could act as toxicant (Naqvi *et al.*, 1989). They revealed a moderate to high level of inhibition of both nucleic acids. Phillips and Loughton (1979) also reported inhibition in RNA and protein synthesis. Sixty per cent RNA inhibition was found after dimilin treatment (Perveen, 2012). In the light of above findings and the present experimental data it may be concluded that fenoxycarb acts as a toxicant and interferes with the synthesis sites of DNA and RNA in the testes as well as ovaries and hence inhibits the synthesis of both the nucleic acids resulting into its reduced level in the gonadial tissues of *C. cephalonica*.

Fenoxycarb exposure caused a significantly dose-dependent ($p < 0.01$) reduction in the levels of DNA (Table 18, Figure 31) and RNA (Table 19, Figure 32) in the eggs, laid within 24 h of mating by different crosses involving one or the other or both sexes emerged from the treated culture. The degree of reduction was found to be maximum in the eggs obtained from the crosses between treated males and treated females and minimum in the eggs of the crosses between normal females and treated males. Similar result has also been observed in the eggs of *S. litura* exposed to an IGR, chlorfluazuron (Perveen, 2012).

In the present investigation, fenoxycarb toxicity decreased the total protein level in the eggs as well as the DNA, RNA and total protein levels in the ovaries of *C. cephalonica*. Thus, it may be assumed that fenoxycarb interferes with the synthesis sites of nucleic acids in gonadial tissues and thus reduce lavel of DNA and RNA in the eggs.

Carbohydrates are one of the most essential biochemical constituents of insect tissues, many of which support optimum growth, development, reproductive activity and survival of individual species (Chefurka, 1959, 1965a, 1965b; Kilby, 1963; Wyatt, 1967 and Friedman, 1970).

In the present investigation, fenoxycarb treatment revealed a significantly dose-dependent ($p < 0.05$) reduction in the glycogen content of testes and ovaries of the adults emerged from the treated culture (Table 20, Figure 33). This reduction was observed to be increasing with the increase in the concentration of the fenoxycarb in the rearing medium. The observations on glycogen level in testes and ovaries clearly indicate that it reduces with the increase in the concentration of methoprene, which may be directly related with the reduction in the glycogen content of the newly laid eggs obtained from the treated parents.

Similar findings have also been reported by other workers in the gonadial tissues of insects of different orders following exposure of IGRs like FMC-23509 (CRD-9499), a juvenile hormone analogue, on *G. gryllotalpa* (Mandal, 1982); tebufenozide, an agonist of moulting hormone, on *E. kuehniella* (Khebbeb *et al.*, 2008) and chlorfluazuron, a chitin synthesis inhibitor, of *S. litura* (Perveen, 2012). FMC-23509 (CRD-9499) caused a significant reduction in the glycogen levels in testes and ovaries of *Gryllotalpa gryllotalpa* (Mandal, 1982). Khebbeb *et al.* (2008) reported that application of tebufenozide caused a significant decrease in carbohydrate content in testes and ovaries of *E. kuehniella*. An insignificant reduction in carbohydrate content of ovaries and a significant decrease in carbohydrate content of testes has also been reported in case of *Spodoptera litura* following exposure of chlorfluazuron

(Perveen, 2012). Methoprene, a juvenile hormone analogue, when applied to fifth instar larvae of *B. mori* resulted into significant increase in glycogen and trehalose in their haemolymph and fat body (Begum *et al.*, 2011).

As to the effect of corpora-allata on carbohydrates metabolism, Morohoshi and Kiguchi (1969), Odhiambo (1966), Liu (1973, 1974), Choudhuri and Mandal (1981) and Mandal (1982) had already reported that allatectomy of newly emerged adult insect insects resulted in accumulation of glycogen in different tissues. The present investigation confirms the role of JH on the glycogen content, because when JHA was applied, the glycogen content decreased significantly. So, it might be possible that JH induced the glycogen release from the tissue by acting as hyperglycemic agent or by inducing the rapid mobilization of glycogen. Another possibility might be that the JHA initiated or inhibited the brain or other endocrine portion of the insects which caused this type of effects (Mandal, 1982). It may be presumed that the reduction in the glycogen content may create energy crisis, which may cause inactivation of spermatogonia and spermatocytes as well as may affect spermatogenesis. The reduction in the level of glycogen, in the present investigation in the testes of moths emerged from the treated cultures may be due to inhibited synthesis and/or inhibited storage of glycogen.

Glycogen is hydrolyzed to free carbohydrates like glucose for synthesis of the disaccharide trehalose, which is then released into the haemolymph (Wyatt, 1967; Keeley, 1978; Goldsworthy and Gade, 1983; Candy, 1985; Candy *et al.*, 1997 and Thompson, 2003). Glycogen, the major fat body carbohydrate reserve, is mobilized for use by other tissues, mainly in the form of trehalose, the blood sugar of insects (Thompson, 2003). Glycogen can be synthesized also in the ovary itself, from haemolymph trehalose as major source (Yamashita and Hasegawa, 1985). Engles (1966) and Engles and Bier (1967) have reported by applying H^3 histidine at various stages of vitellogenesis in *Apis mellifica* and *M. domestica* L., that the carbohydrate incorporation occurred primarily during later stages of oocyte maturation. They are of the opinion that the glycogen synthetase, essential for glycogen synthesis, does exist in inactive form in the earlier stages of oocyte maturation. With the help of radioautography, Ramamurty (1968) has also noticed almost similar findings and proposed that the nurse cells and follicle cells of ovaries of scorpion fly, *Panorpa communis* do not supply glycogen to the oocyte. But follicle cells have reported to be the source of glycogen in the ovaries of *Bombus* (Palm, 1948) and honey bee (Bier, 1954). Engles and Drescher (1964) have observed that the labeled carbohydrate appeared in the oocyte within three minutes of application of H^3 glucose, in case of *A. mellifica*. This rapid uptake of glucose further suggests that it is the oocyte itself which is involved in carbohydrate synthesis. Blood glucose has also been reported to be utilized in the glycogen synthesis by the oocytes (Bonhag, 1956 and Ramamurty, 1968).

The glycogen content of the eggs, in the present investigation, also reduced dose-dependently ($p < 0.05$) with the increasing concentrations of fenoxycarb. It was further observed that this reduction was maximum in the eggs collected from the crosses between treated males and treated females, whereas minimum reduction was recorded in the eggs collected from the crosses between treated males and normal females (Table 21, Figure 34).

It may be concluded that, in the present investigation, fenoxycarb affects the synthesis of glycogen in the oocytes by hampering the activity of glycogen synthetase and/or by blocking the passage of raw materials for glycogen synthesis to the oocytes which results in the reduced levels of glycogen in the ovaries and eggs as well. Similar opinion has been suggested by Engles and Bier (1967) for the reduction of glycogen synthesis by radioisotopic studies in *Apis mellifica* and *M. domestica* L.

Acid phosphatase plays a significant role in catabolism, pathological necrosis, autolysis and phagocytosis (Conchie *et al.*, 1959; De Duve, 1959; Von Lacker and Holtzer, 1959; Becker and Barron, 1961; Abou-Donia, 1978 and Aruna *et al.*, 1979). It has also been reported to have proteolytic action (Bell *et al.*, 1970 and Moczon, 1976) and helps in energy liberating processes (Dalela *et al.*, 1978). Alkaline phosphatase has been reported to be involved in the transport of metabolites across the cell membranes (Danielli, 1952 and Vorbrodt, 1959), synthesis of certain enzymes (Sumner, 1965), protein synthesis (Sumner, 1965; Pilo *et al.*, 1972 and Shaffi and Jafri 1974), secretory activity (Ibrahim *et al.*, 1974) and spermatogenesis (Pavlikova and Repas, 1975), hydrolysis of phosphomonoesters under the alkaline condition (Miao, 2002). Both the enzymes (acid and alkaline phosphatases) have been shown to be associated with insect development, especially in relation to nutrition and egg maturation (Tsumuki and Kanehisa, 1984).

In the present study, it was observed that sub lethal concentrations of fenoxycarb caused a significantly dose-dependent ($p < 0.05$) enhancement in the activity of acid phosphatase and a significantly dose-dependent ($p < 0.05$) reduction in the activity of alkaline phosphatase in the tissues of the gonads *i.e.* testes and ovaries of the moths emerged from the treated culture (Table 22, Figures 35 and 36).

Similar to the present investigation, JHA *i.e.* FMC-23509 (CRD-9499) induced significant enhancement in the activity of acid phosphatase has also been reported in the testes and ovaries of *G. gryllotalpa* (Mandal, 1982). Saha *et al.* (1985) has also reported that application of JHA caused a significant increase in the activity of acid phosphatase and a significant decrease in the activity of alkaline phophatase in both the tissues *i.e.* testes and ovaries of *C. stolli*. Effect of five insect growth regulators (IGRs) *i.e.* buprofezin, hexaflumuron and lufenuron, which are chitin synthesis inhibitors (CSIs); tebufenozide, an ecdysone agonist (EA) and pyriproxyfen, a juvenile hormone analogue (JHA), on the activity of acid phosphatase was reported against the whole larva of housefly *M. domestica* (Assar *et al.*, 2010). The tested IGRs significantly increased the activity of acid phosphatase except tebufenozide which exhibited a severe (significant) reduction in the enzyme activity as compared to the control (Assar *et al.*, 2010). Hexaflumuron gave the highest increase in acid phosphatase activity followed by lufenuron, buprofezin and pyriproxyfen. In addition, cyromazine, a chitin synthesis inhibitor (CSI), also caused a significant enhancement in the activity of acid phosphatase and a significant reduction in the activity of alkaline phosphatase in fourth instar larva of *Culex pipiens*, as compared to the control (Assar *et al.*, 2012). Similarly, IGRs induced increase in acid phosphatase activity have also been reported by different workers in various insects such as,

20-hydroxy ecdysone on *Manduca sexta* (Caglayan, 1990); pyriproxyfen on *C. pipiens* (El-Bassal, 1993); pyriproxyfen on *Pectinophora gossypiella, E. insulana* (Anan *et al.*, 1993); *S. littoralis* (Abdel-Aal, 2002) and on *A. ipsilon* (El- Sheikh, 2002). JHA treatment also caused increase in soluble and membrane bound activity of acid phosphatase in mid gut, hind gut and malpighian tubules of Colorado potato beelte, *L. decemlineata* (Yi and Adams, 2001). On the basis of above findings and present investigation, it may be concluded that fenoxycarb accelerates the activity of acid phosphatase which may interfere with the gametogenesis in gonads and may also check the further development of the zygote by reducing the protein level in the gonadial cells or by other catabolic activities. Acid phosphatase has received considerable attention in developmental studies because of its association with histolysis. It is known that acid phosphatase hydrolyzes a variety of orthophosphorylation reactions (Hollander, 1971).

JH induced increased activity of acid phosphatase has also been reported in testes and ovaries of *G. gryllotalpa* (Mandal, 1982). He suggested that increased activity of acid phosphatase in gonads of *G. gryllotalpa* might be due to JH induced cell proliferation of the gonads. Similar reason has also been suggested for the enhancement of acid phosphatase activity in gonads of milkweed bug, *Oncopeltus fasciatus* following exposure of juvenile hormone (Beel and Feir, 1977). Our findings regarding phosphatase activity are in accordance with the results of Mandal (1982) as observed in testes and ovaries of *G. gryllotalpa* following exposure of a juvenile hormone analogue *i.e.* FMC-23509 (CRD-9499); Saha *et al.* (1985) as observed in testes and ovaries of *C. stolli* exposed to JHA; Assar *et al.* (2012) in the larva of *C. pipiens* treated with cryomazine which is a chitin synthesis inhibitor; Caglayan (1990) in case of *M. sexta* exposed to 20-hyroxyecdysone; El-Bassal (1993) in *C. pipiens*, Anan *et al.* (1993) in *P. gossypiella* and *E. insulana*, Abdel- Aal (2002) in *S. littoralis* and El- Sheikh (2002) in *A. ipsilon* exposed to pyriproxyfen. On the basis of above findings and present investigation, it may be concluded that enhancement in acid phosphatase activity in testes and ovaries of *C. cephalonica* may be due to methoprene induced cell proliferation of the gonads.

Etebari *et al.* (2007) investigated that alkaline phosphatase activity of haemolymph in treated and control group of silkworm larvae in 24 and 120 h after JHA application, in order to see the enzyme activity through the pyriproxyfen mode of action. Some other IGRs like pyriproxyfen has also been reported to reduced the activity of alkaline phophatase in *A. ipsilon* (El- Sheikh, 2002) while the activity of this enzyme was not affected by the same compound in *C. pipiens* (El-Bassal, 1993). Sridhara and Bhat (1963) stated that the increase or decrease of both phosphatase enzymes during development is reflected by an increase or decrease in the acid-soluble phosphorus content.

Reduced activity of alkaline phosphatase cause agglutination of sperms in the testes and is likely to be associated with the mitochondria and therefore would lead to loss of sperm motality (Campion, 1972). Thus, it may be inferred that the reduced activity of alkaline phosphatase, in the testes, due to methoprene treatment, may interfere with the activities of sperms and thereby causing sterility in the male individuals of *C. cephalonica*. Kawatra (1976) has mentioned that alkaline phosphatase is necessary for the ovarian development. Therefore, it may be supposed that the inhibition of alkaline phosphatase activity, in the ovaries of this moth, may be one of the possible factors for the inhibition of ovarian development. Thus, it may be concluded that the decrease in the activity of alkaline phosphatase might be the result of direct and/or indirect involvement of methoprene in the inhibition of its activity or its synthesis or both.

The sub lethal concentrations of fenoxycarb also revealed a significantly dose-dependent ($p < 0.05$) enhancement in the activity of acid phosphatase and a similar associated dose-dependent reduction in the activity of alkaline phosphatase in the eggs obtained from different crosses. The maximum enhancement in the activity of acid phosphatase and maximum reduction in the activity of alkaline phosphatase was observed in the eggs collected from the crosses between treated males and treated females, whereas the minimum enhancement in the activity of acid phosphatase and minimum reduction in the activity of alkaline phosphatase was recorded in the eggs collected from the crosses between treated males and normal females (Table 23 and 24 Figures 36 and 38).

It may be concluded that methoprene induced enhancement in the activity of acid phophatase and reduction in the activity of alkaline phosphatase in the eggs of *C. cephalonica* may be related with the abnormalities of both the phosphatase enzymes of ovaries of this insect exposed to methoprene. The variations in the acid and alkaline phosphatase activities in ovaries, due to methoprene action in the present investigation, might lead to the elevated activity of acid phosphatase and declined activity of alkaline phosphatase in the eggs of *C. cephalonica*.

Chapter 5

Summary

The present study, for the first time, reveals some of the so far unexplored fenoxycarb induced changes on the ontogeny (larva, pupa and adults); growth duration and longevity of adults and reproductive behaviour (fecundity and fertility) of the rice moth *Corcyra cephalonica* exposed as first, second, third and fourth instars larvae. The biochemical changes, caused by fenoxycarb, in the levels of total protein, total free amino acids, nucleic acids and glycogen and in the activities of acid and alkaline phosphatases in the testes and ovaries as well as in the eggs laid within 24 h of mating, have also been assessed.

The different larval stages (first, second, third and fourth instars), of *C. cephalonica* were continuously exposed till the adult emergence from the diet containing different grades of fenoxycarb. On the completion of life-cycle, per cent adult emergence and percent pupal mortality was observed and on that basis percent pupation and percent larval mortality was calculated. Some of the emerged adults were allowed to mate, while others were dissected for obtaining the gonads for biochemical estimations. From the mated females, eggs were collected to study the egg laying capacity of female and egg hatchability. The eggs laid within the 24 h of mating were used for biochemical estimations. All the experiments were performed at temperature 26 ± 1°C, relative humidity (R.H.) 93 ± 5 per cent and a light regime of 12 h light and 12 h darkness.

The larval and pupal mortalities were found to be increased with the increase in concentration of fenoxycarb, and hence decreasing the per cent adult emergence in all the instars of *C. cephalonica*. Younger larvae were found to be more susceptible to fenoxycarb treatment and the concentrations above 0.10 ppm were recorded to be highly lethal at different life stages of this insect and at 0.50 ppm and 1.00 ppm concentrations of fenoxycarb there was 100 per cent suppression of adult emergence in all the instars.

The higher concentrations of this compound severely disrupt the metamorphosis of *C. cephalonica* and abnormal adults and abnormal pupae were produced. In case of 1st instar treated larvae some abnormal adults were produced at the concentrations of 0.01, 0.05 and 0.10 ppm of fenoxycarb while in case of second, third and fourth instar treated larvae abnormal adults were formed but in decreased fashion from first instar to fourth instar. Degree of abnormalities ranges from no morphological distinguishing clue for male and female, unfolded or twisted wings, twisted legs and abnormally long abdomen in males and too much swollen abdomen in females. It also deserves mention that by increasing the concentrations of the fenoxycarb also resulted in the formation of abnormal pupae or larval pupal intermediates. In case of first and second instar treated larvae, at 0.05, 0.10 and 0.50 ppm concentrations of fenoxycarb abnormal pupae were formed, while in case of third and fourth instars treated larvae abnormal pupae were formed at the concentrations of 0.10, 0.50 and 1.00 ppm concentrations of fenoxycarb.

In addition, the higher concentrations of fenoxycarb *i.e.*0.01, 0.05, 0.10, 0.50 and 1 ppm produced giant larvae, supernumerary larvae, larval-pupal intermediates It deserves mention that in case of first and second instar treated larvae at 0.50 and 1 ppm concentrations, there was neither the emergence of any adult nor any adultoid but extra moult of larvae was observed resulting into the formation of supernumerary and giant larvae which after certain tenure of life get shrinked and thenafter became blackish and ultimately died.

At increasing concentrations of fenoxycarb, in all the instars, the growth duration of *C. cephalonica* was observed to be increased compared to the controls. It deserves mention that longevity of the adults was found to be decreased with increasing concentration of fenoxycarb and this effect was found to be maximum on the adults emerged from the first instar larval treatments.

The egg laying capacity (fecundity) of the female reduced significantly with the increase of fenoxycarb concentration. Maximum reduction in the number of eggs laid per female was observed in the crosses between treated males and treated females and minimum in the crosses between normal females and treated males, which shows that females are more susceptible than the males. The inhibition in the egg laying capacity of the females was also observed to be directly related with the concentration of fenoxycarb, age of larval treatment and duration of fenoxycarb exposure to the insect.

The egg hatchability (fertility) was also found to be reduced with increasing concentration of fenoxycarb and it was observed that maximum reduction was in eggs collected from the crosses between treated males and treated females, whereas minimum in the crosses between normal females and treated males. It further reflects more susceptibility of females than males.

For biochemical estimations, out of various concentrations of the fenoxycarb mentioned above, only such concentrations (0.001, 0.005 and 0.01 ppm) were selected, which allowed the larvae to survive and develop but caused considerable effect in the internal biochemistry of the gonads (testes and ovary) and eggs (laid within 24 h of mating) that could be easily detected and assessed to prove the effectiveness of hormonal control measures against this lepidopterous pest. Since first instar larvae were more susceptible to fenoxycarb treatment and their tenure of exposure was also much long, therefore, only first instar larvae treated with the above mentioned sub lethal concentrations were taken into consideration for biochemical studies.

Sublethal concentrations of the fenoxycarb caused a significantly dose-dependent reduction in the levels of total protein, DNA, RNA, glycogen and activity of alkaline phosphatase and a significantly dose-dependent enhancement in the levels of total free amino acids and activity of acid phophatase in the gonadial tissues as well as in the eggs of *C. cephalonica* compared to controls.

The degree of inhibition in the levels of total protein, DNA and RNA, glycogen and the activity of alkaline phosphatase and enhancement in the total free amino acids level and the activity of acid phosphatase was found to be maximum in the eggs collected from the crosses between treated males and treated females, and minimum, in the eggs collected from the crosses between normal females and treated males. It again confirms the view that the effect of fenoxycarb was more in females than the males of *C. cephalonica*.

The effect of fenoxycarb on different biochemical constituents of testes may cause various deformities such as distortion of basement membrane, pycnosis of spermatogonia, degeneration of the primary as well as secondary spermatocytes and dissociation of spermatozoa which may finally lead to the reduced and malformed development of sperms. The biochemical changes in the ovaries, caused by fenoxycarb treatment, may affect the normal functioning of nurse cells, follicular epithelium, oogenesis and developing oocytes and ultimately may lead to the decreased and malformed production of ova in females. The changes in the biochemical constituents of gonads might be, in turn, reflected in eggs of different crosses and hence it is assumed that their fecundity and fertility is severely affected.

It is noticeable that in case of first instar treated larvae the duration of fenoxycarb exposure to the insect was much longer in comparison to second, third and fourth instar treated larvae, thus, the influence of this JHA to mortality at post-embryonic stages, longevity of adults, fecundity and fertility as well as gonadial and egg biochemistry was found to be much pronounced in case of first instar treated larvae in comparison to first, second and third instars treated. On the basis of the present investigation it is obvious that toxicity of fenoxycarb is directly related with its concentration, age of the larval treatment, sex of the insect and the duration of fenoxycarb exposure.

Thus, the results obtained in the present investigation *i.e.* increased mortality at post-embryonic stages, decreased longevity of adults, reduction in the egg laying capacity and egg hatchability, as well as the information obtained with special reference to changes in the biochemical framework including enzyme activity of gonads and eggs (laid within 24 h of mating), if employed, would be of immense importance and be able to reduce the population build up of *C. cephalonica* in stored cereals. This knowledge will also facilitate in framing pest management programmes in future. It also deserves mention that for the effective control of the rice moth, *C.*

cephalonica application of fenoxycarb should be done during initial stage of its life-cycle, so that evolution of a new generation of this pest for the eventual establishment on stored cereals and cereal products can be considerably restricted.

References

Abbott, W.S. 1925. A method of computing the effectiveness of an insecticide. *J. Econ. Entomol.*, 18: 265-267.

El-Mageed, A.F.N. 2008. Physiological effects of certain bioagents and an insect growth *Spodoptera littoralis* (Bosid.) (Noctuidae: Lepidoptera). M.Sc. Thesis, Fac. of Agric., Ain Shams Univ., Cairo, Egypt.

Abdel-Aal, A.E. 2002. Effect of some insect growth regulators on certain biological, biochemical and histopathological aspects of the cotton leafworm, *Spodoptera littoralis* (Bosid.) (Lepidoptera: Noctuidae). Ph.D. Thesis, Faculty of Science, Cairo Univercity, Giza, Egypt.

Abou-Donia, M.B. 1978. The role of acid phosphatase in delayed neurotoxicity induced by leptophos in hens. *Biochem. Pharmacol.*, 27: 2055-2058.

Adan, A., P.D. Estal, F. Budia, V. Narco, J. Jacas and E. Vinuela, 1994. Effects of hexaflumuron on the fecundity and fertility of *Tribolium castaneum* Herbst (Coleoptera: Tenebrionidae). *Bol. Sanid. Veg. Plagas*, 20(2): 371-377.

Ahi, J. and S. Rai, 2012. Effect of fenoxycarb, a juvenile hormone mimic on the testes of *Poekilocerus pictus* (F.). *Indian J. Entomol.*, 74(2): 142-147

Alizadeh, M., J. Karimzadeh, G.R. Rassoulian, H. Farazmand, V. Hoseini-Naveh and H.R. Pourian, 2012. Sublethal effects of Pyriproxyfen, a Juvenile Hormone Analogue, on *Plutella xylostella* (Lepidoptera: Plutellidae): Life table study. *Arch. Phytopathol. Plant Protect.*, 45(14): 1741-1763.

Ambika, B. and C.C. Abraham, 1982. Effect of juvenile hormone analogue, methoprene on development of eggs and larvae of *Corcyra cephalonica* Stainton (Lepidoptera: Pyralidae). *Kerla Agril. Res. J.*, 20: 60-62.

Amiri, A., R. Bandani and A. Darvishzadeh, 2012. Effects of the insect growth regulators methoxyfenozide and pyriproxyfen on adult diapause in sunn pest *Eurygaster integriceps* (Hemiptera: Scutelleridae). *J. Agr. Sci. Tech.*, 14: 1205-1218.

Ammar, I.M.A. 1990. Biological activity of certain IGRs against the immature stages of the rice weevil, *Sitophilus oryzae* L. (Coleoptera: Curculionidae). *Zagazig J. Agric. Res.*, 17: 465-470.

Amos, T.G., P. Williams and R.L. Semple, 1978. Sterilizing activity of methoprene and hydroprene in *Tribolium castaneum* (Herbst). *Experientia*, 34(4): 469-470.

Anan, A. R., I. M. Mona and M. H. Nagwa, 1993. Biochemical effect of pyriproxyfen juvenoid on fat and haemolymph proteins of pink bollworm, *Pectinophora gossypiella* (Saund.) and spiny bollworm, *Earias insulana* (Boisd.). *Ann. Agric. Sci.*, 38: 761-772.

Andersch, M.A. and A.J. Szcypinski, 1947. *Am. J. Clin. Pathol.*, 17: 571, and In: Methods in Enzymatic Analysis. Bergmeyer, U.H. (ed.), Academic Press, New York, 1967.

Anonymous, 1973. Where are the markets? Farm Chem. *Crop Life*, 136(1): 26-30.

Arbogast, R.T., P.E. Kendra, R.W. Mankin and R.C. McDonald, 2002. Insect infestation of a botanicals warehouse in north-central Florida. *J. Stored Prod. Res.*, 38: 349-363.

Aribi, N., G. Smagghe, S. Lakbar, N. Soltani-Mazouni and N. Soltani, 2006. Effects of pyriproxyfen, a juvenile hormone analog, on development of the mealworm, *Tenebrio molitor*. *Pestic. Biochem. Physiol.*, 84: 55-62.

ARS Pesticide Properties Database, USDA, ARS, Remote Sensing and Modeling Lab, Beltsville, Maryland, 1995.

Arthur, F.H. 1996. Grain Protectants: Current Status and Prospects for the Future. *J. Stored Prod. Res.*, 32(4): 293-302.

Arthur, F.H. 2001. Susceptibility of last instar red flour beetles and confused flour beetles (Coleoptera: Tenebrionidae) to hydroprene. *J. Econ. Entomol.*, 94(3): 772-779.

Arthur, F.H. 2003. Efficacy of a volatile formulation of hydroprene (PointsourceTM) to control *Tribolium castaneum* and *Tribolium confusum* (Coleoptera: Tenebrionidae). *J. Stored Prod. Res.*, 39: 205-212.

Arthur, F.H. and T.W. Phillips, 2003. Stored-Product Insect Pest Management and Control. In: Food Plant Sanitation. Hui, Y.H., B.L. Bruinsma, J.R. Gorham, W.K. Nip, P.S. Tong and P. Ventresca (eds.), *Marcel Dekker*, New York, 341-358.

Arthur, F.H., S. Liu, B. Zhao and T.W. Phillips, 2009. Residual efficacy of pyriproxyfen and hydroprene applied to wood, metal and concrete for control of stored-product insects. *Pest Manag. Sci.*, 65(7): 791-797.

Arthur, F.H. 2010. Residual efficacy of aerosols to control *Tribolium castaneum* and *Tribolium confusum*. *Proc. Stored Prod. Protect. Int. Working. Conf.* Estoril, Portugal. pp. 789-792.

Arthur, F.H. and E.A. Fontenot, 2012. Residual activity of methoprene and novaluron as surfacetreatments to manage the flourbeetles, *Tribolium castaneum* and *Tribolium confusum*. *J. Insect Sci.*, 12: 95.

Aruna, P., C.S. Chetty, R.C. Naidu and K.S. Swami, 1979. Acid phosphatase activity in Indian apple snail, *Pila globosa* (Swainson) during aestivation and starvation stress. *Proc. Indian Acad. Sci.*, 88 B: 363-365.

Ashok, M., C. Turner and T.G. Wilson, 1998. Insect juvenile hormone resistance gene homology with the bHLH-PAS family of transcriptional regulators. *Proc. Natl. Acad. Sci.*, 95: 2761-2766.

Ashok, V. 2000. Endocrine regulation of spermatogenesis in *Spilostethus pandurus* (Scopoli) (Hemiptera: Lygaeidae). *J. Exp. Zool.*, 3: 119-125.

Assar, A.A. 2004. Histochemical effects of some insect growth regulators on the larvae of flesh fly, *Parasarcophaga aegyptiaca* Salem (Diptera: Sarcophagidae). *J. Egypt Acad. Soc. Environ. Dev.*, 5(2): 73-89.

Assar, A.A. and A.A. Abo-Shaeshae, 2004. Effect of two insect growth regulators, methoxyfenozide and pyriproxyfen on the housefly, *Musca domestica vicina* (Diptera: Muscidae). *J. Egypt. Ger. Soc. Zool.*, 44(E): 19-42.

Assar, A.A., M.M. Abo EL-Mahasen, M.E. Khalil and S.H. Mahmoud, 2010. Biochemical effects of some insect growth regulators on the houses fly, *Musca domestica* (Diptera: Muscidae). *Egypt. Acad. J. Bio. Sci.*, 2 (2): 33-44.

Assar, A.A., M.M. Abo-El-Mahasen, N.M. Harba and A.A. Rady, 2012. Biochemical effects of cyromazine on *Culex pipiens* larvae (Diptera: Culicidae). *J. American Sci.*, 8(5): 443-450.

Athanassiou, C.G., F.H. Arthur and J.E. Throne, 2010. Efficacy of methoprene for control of five species of psocids (Psocoptera) on wheat, rice, and maize. *J. Food Prot.*, 73: 2244-2249.

Athanassiou, C.G., F.H. Arthur and J.E. Throne, 2011a. Efficacy of layer treatment with methoprene for control of *Rhyzopertha dominica* (Coleoptera: Bostrychidae) on wheat, rice and maize. *Pest. Manag. Sci.*, 67: 380-384.

Athanassiou, C.G., F.H. Arthur, N.G. Kavallieratos and J.E. Throne, 2011b. Efficacy of pyriproxyfen for control of stored-product psocids (Psocoptera) on concrete surfaces. *J. Econ. Entomol.*, 104(5):1765-1769.

Attri, B.S. and G.R. Prasad, 1980. Studies on the pesticidal value of neem oil by-products. *Pestology*, 4: 16-20.

Attri, B.S. and R. Prasad, 1980a. Neem oil extraction: An effective mosquito larvicide. *Indian J. Entomol.*, 42(3): 371-374.

Atwal, A.S. 1976b. Agricultural Pests of India and South-East Asia. *Kalyani Publishers*, Delhi, pp: 502.

Ayyar, T.V.R. 1919. Some insects recently noticed as injurious in South India. In: Report of the Proceedings of the Third Entomological Meeting. Fletcher, T.B. (ed.), Vol. 1, *Pusa, Superintendent Government Printing*, Calcutta, pp. 314-328.

Ayyar, P.N.K. 1934. A very destructive pest of stored in South India, *Corcyra cephalonica*, Staint. (Lep.). *Bull. Entomol. Res.*, 25(2): 155-169.

Baker, J.E. and P.T. Lum, 1976. Effects of dietary methoprene on symbiotic and asymbiotic rice weevils and aposymbiotic granary weevils. *J. Georgia Ent. Soc.*, 11(3): 213-216.

Bakr, R.F.A. 1986. Morphogenic and physiological aberration induced by certain IGRs in the house fly, *Musca domestica*, Ph.D. Thesis, Faculty of Science, *Ain Sham* University, Cairo, Egypt.

Bakr, R.F.A., N.A. Abdel-Razak, M.S. Hamed and A.M. Guneidy, 1989. Physiological effect of some insect growth regulators on the respirometric movements, total protein and free amino acids of the housefly, *Musca domestica* (L.). *Ain Shams Sci. Bull.*, 28: 169-183.

Bakr, R.F.A., H.M. Abdel-Fattah and E.K. Mohamed, 2007. Effect of chitin synthesis inhibitor, lufenuron on the development, haemolymph and antennal sensilla of *Schistocerca gregaria* (Forskal) (Orthoptera: Acrididae). *African J. Biol. Sci.*, 3(2): 35-43.

Baldessari, M., F. Trona and G. Angeli and C. Ioriatti, 2010. Effectiveness of five insecticides for the control of adults and young stages of *Cacopsylla melanoneura* (Forster) (Hemiptera: Psyllidae) in a semi-field trail. *Pest Manag. Sci.*, 66(3): 308-312.

Beadle, L. and J. Shaw, 1950. The retention of salt and the regulation of the non-protein nitrogen fraction in the blood of the aquatic larva, *Sialia lutaria. J.Exp. Biol.*, 27: 96-109.

Becker, N.H. and K.D. Barron, 1961. The cytochemistry of anoxic and anoxic-ischemic encephalopathy in rats. *Am. J. Pathol.*, 38(2): 161-175.

Begum, A.N., S.M. Moorthy, S. Venkat, S.N. Kumar and S.M.H. Qadri, 2011. Influence of juvenile hormone analogue, methoprene on the biochemical changes and economic characters of silkworm *Bombyx mori*. I. *J. Plant Anim. Environ. Sci.*, 1(3): 171-178.

Bell, G.H., J.N. Davidson and Smith, D.E., 1970. In: Text book of Physiology and Biochemistry, 8th edition, Tinling and Co. London and Brescot. pp. 103-105.

Bell, H.A. and J.P. Edwards, 1999. The activity of (S)-hydroprene space spray against three stored products pests in a simulated food production environment. *J. Stored Prod. Res.* 5(2): 117-126.

Bengston, M. 1986. Insect growth regulators. In: Proceding of Fourth International Working Conference on Stored Production Protection. Donahaye, E. and S. Navarro (eds.), *Tel Aviv*, Israel, pp. 35-46.

Benny, T.M. and V.S.K. Nair, 2006. Effects of juvenile hormone analogue and ecdysone agonist on the spermatogenesis in *Spodoptera mauritia* Boisd. (Lepidoptera: Noctuidae): a flow-cytometric study. *Entomon*, 31(2): 99-106.

Bhargava, M.C. and K.C. Devraj Urs, 1992. Activity of juvenile hormone analogue (RO 20-3600) on larvae of *Corcyra cephalonica* (Staint.) (Lepidoptera: Pyralidae). *Bull. Grain Tech.*, 30(2): 119-124.

Bhargava, M.C. and K.C. Devaraj Urs, 1993. Ovicidal effect of three insect growth regulators on *Corcyra cephalonica. Indian J. Plant Prot.*, 21: 195-197.

Bhatt, R.S. and S.S. Krishna, 1980. Carbohydrases in the developing eggs of the rice-moth, *Corcyra cephalonica* Staint. *Z. Angew. Zool.*, 67: 319-322.

Bhatt, R.S. and S.S. Krishna, 1982. Changes in the content of some biochemical components during development of eggs of the rice-moth, *Corcyra cephalonica* Staint. *Zool. Mag.*, 31: 70-73.

Bhatt, R.S. and S.S. Krishna, 1984a. Effect of some nutritional factors on the free amino acids and protein content of larval haemolymph and fat body of *Corcyra cephalonica. J. Adv. Zool.*, 5(2): 90-101.

Bhatt, R.S. and S.S. Krishna, 1984b. A study of some changes in egg chemistry during embryogenesis in *Dysdercus koenigii* (F.). *Soc. Paper No.*, 927: 3-11.

Bhatt, R.S. and S.S. Krishna, 1986. Effect of some nutritional manupulations on glucose and trehalose levels in larval haemolymph and fat body of *Corcyra cephalonica* (Stainton) (Lepidoptera: Pyralidae). *Entomon*, 11(3): 161-163.

Bier, K. 1954. Uber Phasen gesteigerter protein und kohlenhydrateinlagerung und die fettverteiling in Hymenopternovar. *Zool. Anz. Suppl.*, 18: 422-429.

Billas, I.M.L., L. Moulinier, N. Rochel and D. Moras, 2001. Crystal structure of the ligand-binding domain of the ultraspiracle protein USP, the ortholog of retinoid X receptors in insects. *J. Biol. Chem.*, 276: 7465-7474.

Bishop, G.H., A.P. Briggs and E. Ronzoni, 1925. Body fluids of honey bee larva: II. Constituent of blood, and their osmotic effects. *J. Biol. Chem.*, 66: 77-88.

Blaine, W.D. and S.E. Dixon, 1970. Hormonal control of spermatogenesis in the cockroach, *Periplaneta americana* (L.). *Can. J. Zool.*, 48(2): 283-287.

Bodenstein, D. and I.B. Sprague, 1959. The developmental capacities of the accessory sex glands in *Periplaneta americana. J. Exp. Zool.*, 142(1):177-202.

Bonhag, P.F. 1956. The origin and distribution of periodic acid-Schiff-positive substances in the oocyte of the earwig, *Anisolabis maritima* (Gene). *J. Morphol.*, 99(3): 433-463.

Bortolotti, L., C. Porrini, A.M. Sbrenna and G. Sbrenna, 2000. Ovicidal action of fenoxycarb on a predator, *Chrysoperla carnea* (Neuroptera: Chrysopidae). *Appl. Entomol. Zool.*, 35 (2): 265-270.

Bortolotti, L., A.M. Sbrenna and G. Sbrenna, 2005. Action of fenoxycarb on metamorphosis and cocoon spinning in *Chrysoperla carnea* (Neuroptera: Chrysopidae): identification of the JHA-sensitive period. *Eur. J. Entomol.*, 102(1): 27-32.

Bownes, M. 1989. The roles of juvenile hormone, ecdysone and the ovary in the control of *Drosophila vitellogenesis. J. Insect Physiol.*, 35(5): 409-413.

Brachet, J. 1955. In: The Nucleic Acids. Vol. II, Academic Press, New York

Brindle, P.A., F.C. Baker, L.W. Tsai, C.C. Reuter and D.A. Schooley, 1987. Sources of proprionate for the biogenesis of ethyl-branched insect juvenile hormones: role of isoleucine and valine. *Proc. Natl. Acad. Sci. U.S.A.*, 84(22): 7906-7910.

Brown, M.R. and J.J. Brown, 1982. Effect of methoprene on the fecundity and fertility of the codling moth *Cydia pomonella*. *Ann. Entomol. Soc. Am.*, 75(3): 257-260.

Buck, J.B. 1953. The internal environment in regulation and metamorphosis. In: Insect Physiology. Roeder, K. (ed.), *Wiley*, New York, pp. 191-217.

Burkholder, W.E. 1990. Practical use of pheromones and other attractants for stored-product insects. In: Behaviour-Modifying Chemicals for Insect Manegment. Ridgway, R.L., R.M. Silverstein and M.N. Inscoe (eds.), *Marcel Dekker Inc.*, New York and Basel, pp. 497-516.

Burkholder, W.E. and D.L. Faustini, 1991. Biological methods of survey and control. In: Ecology and Management of Food Industry Pests. Gorham, J.R. (ed.), *Association of Official Analytical Chemists*, Arlington, Virginia, pp. 361-372.

Burton, S.D. 1995. Aerobic aquatic metabolism of 14C -fenoxycarb. *Ciba-Geigy Corporation Data Package*, Report No. 163859-N. DPR No. 51317-085.

Butenandt, A. and P. Karlson, 1954. Uber die isolierung eines metamorphose-hormons des insekten in kristallisierter from. *Z. Naturforsch B* 9: 389-391.

Butlin, R.K., C.W. Woodhatch and G.M. Hewitt, 1987. Male spermatophore investment increases female fecundity in a grasshopper. *Evolution*, 41(1): 221-225.

Caglayan, S.H. 1990. Effect of 20-hydroxyecdysone on acid phosphatase activity in the larval fat body of *Manduca sexta*. *Biochem. Int.*, 20(3): 511- 518.

Campion, D.G. 1972. Insect chemosterilants: A review. *Bull. Entomol. Res.*, 61: 577-635.

Candy, D.J. 1985. Intermediary metabolism. In: Comprehensive Insect Physiology Biochemistry and Pharmacology. Kerkut, G.A. and L.E. Gilbert (eds.), Vol. 10, *Pergamon Press*, New York, pp. 1-41.

Candy, D.J., A. Becker and G. Wegener, 1997. Coordination and integration of metabolism in insect flight. *Comp. Biochem. Physiol. B: Biochem. Mol. Biol.*, 117(4): 497-512.

Cappellozza, L., S. Cappellozza and G. Sbrenna, 1995. Changes in the developmental programme of *Bombyx mori* caused by oral treatment with fenoxycarb during the last larval instar. *Sericologia*, 35: 427-436.

Carter, D.J. 1975. Laboratory evaluation of three novel insecticides inhibiting cuticle formation against some susceptible and resistant stored products beetles. *J. Stored Prod. Res.*, 11(3-4): 187-193.

Chamberlain, W.F. 1962. Chemical Sterilization of the Screw-Worm. *J. Econ. Entmol.*, 55(2): 240-248

Chanbang, Y., F.H. Arthur, G.E. Wilde, J.E. Throne and B. Subramanyam, 2008. Susceptibility of eggs and adult fecundity of the lesser grain borer, *Rhyzopertha dominica*, exposed to methoprene. *J. Insect Sci.*, 8: 1-5.

Chandra, A. and S.K. Tiwari, 2013. Insecticidal effect of methoprene on the pre-adult stages of almond moth, *Ephestia cautella* Walker (Lepidoptera: Pyralidae). *J. Biol. Erath Sci.*, 3: 269-274.

Chang, T.-H. and J.G. Riemann, 1967. H^3-Thymidine radioautographic study of spermatogenesis in the boll weevil, *Anthonomus grandis* (Coleoptera: Curculionidae). *Ann. Entomol. Soc. Am.*, 60(5): 975-979.

Chapman, R.F. 1982: The Insects: Structure and Function. Third Edition, *Harvard University Press*, Cambridge, MA.

Charles, J.-P., H. Wojtasek, A.J. Lentz, B.A. Thomas, B.C., Bonning and S.R. Palli, 1996. Purification and reassessment of ligand binding by the recombinant, putative, juvenile hormone receptor of the tobacco hornworm, *Manduca sexta*. *Arch. Insect Biochem. Physiol.*, 31(4): 371-393.

Charmillot, P.J., A. Gourmelon, A.L. Fabre and D. Pasquier, 2001. Ovicidal and larvicidal effectiveness of several insect growth inhibitors and regulators on the codling moth *Cydia pomonella* L. (Lep., Tortricidae). *J. Appl. Entomol.*, 125(3): 147-153.

Charmillot, P.J., D. Pasquier, C. Salamin and A. Ter-Hovannesyan, 2007. Ovicidal and larvicidal effectiveness of insecticides applied by dipping apples on the small fruit tortrix *Grapholita lobarzewskii*. *Pest Manag. Sci.*, 63(7): 677-681. doi: 10.1002/ps.1373.

Chefurka, W. 1959. Glucose metabolism in insects. In: Biochemistry of Insects, Symposium XII of the IV Int. Congress Biochem. Levenbook, L. (ed.), *Pergamon Press*.

Chefurka, W. 1965a. Intermediary metabolism of carbohydrates in insects. In: The Physiology of Insects. Rockstein, M. (ed.), Vol. 2, *Academic Press*, New York, pp. 581-667.

Chefurka, W. 1965b. Some comparative aspects of metabolism of carbohydrates in insects. *Ann. Rev. Entomol.*, 10: 345-382.

Chiang, A.-S., W.-Y. Lin, H.-P. Liu, M.A. Pszezalkowski, T.-F. Fu, S.-L. Chiu and G.L. Holbrook, 2002. Insect NMDA receptors mediate juvenile hormone biosynthesis. *Proc. Natl Acad. Sci. U.S.A.*, 99(1): 37-42.

Chinzei, Y. and S. Tojo, 1972. Nucleic acid changes in the whole body and several organs of the silkworm, *Bombyx mori* during metamorphosis. *J. Insect. Physiol.* 18: 1683-1698.

Chittenden, F.H. 1919. The rice moth. U. S. Department of Agricultural, Bulletin No. 783. *Washington (Govt. Print. Off.)*, pp. 15.

Choudhuri, D.K. and S. Mandal, 1981. Fluctuation of glycogen contents in the brain during the post-embryonic development of *Schizodactylus monstrosus* D.; Effect of starvation stress and Corpora allatectomy (Schizodactylidae: Orthoptera). *Indian J. Exp. Biol.*, 19(6): 532-534.

Clayton, G.M., S.Y. Peak-Chew, R.M. Evans and J.W. Schwabe, 2001. The structure of the ultraspiracle ligand-binding domain reveals a nuclear receptor locked in an inactive conformation. *Proc. Natl. Acad. Sci. U.S.A.*, 98(4): 1549-1554.

Cogburn, R.E. 1988. Fenoxycarb as a long term protectant for stored rough rice. *J. Econ. Entomol.*, 81(2): 722-726.

Comas, D., M.D. Piulachs and X. Belles, 1999. Fast induction of vitellogenin gene expression by juvenile hormone III in the cockroach *Blattella germanica* (L.) (Dictyoptera, Blattellidae). *Insect Biochem. Mol. Biol.*, 29: 821-827.

Conchie, J., J. Findlay and G.A. Lievvy, 1959. Mammalian glycosidases. Distribution in the body. *Biochem. J.*, 71: 318-325.

Cotton, R.T. 1956. Pest of stored grains and grain products. *Burgess Publishing Co.*, Minneopolis, MN, p. 318.

Coudron, T.A., J.H. Law and J.K. Koeppe, 1981. Insect Hormones. *Trends Biochem. Sci.*, 6: 248-251.

Cusson, M., C.G. Yu, K. Carruthers, G.R. Wyatt, S.S. Tobe and J.N. McNeil, 1994. Regulation of vitellogenin production in armyworm moths, *Pseudaletia unipuncta. J. Insect Physiol.*, 40(2): 129-136.

Cusson, M. and S.R. Palli, 2000. Can juvenile hormone research help rejuvenate integrated pest management? *Can. Entomol.*, 132(3): 263-280.

Daglish, G.J. and C. Pulvirenti, 1998. Reduced fecundity of *Rhyzopertha dominica* (F.) (Coleoptera: Bostrychidae) following exposure of adults to methoprene. *J. Stored Prod. Res.*, 34: 201-206

Daillie, J. 1979. Juvenile hormone modifies larvae and silk gland development in *Bombyx mori. Biochimie*, 61(2): 275-281.

Dalela, R.C., M.C. Bhatnagar and S.R. Verma, 1978. Histochemical studies on the effect of rogor and thiodon on the activity of acid phosphatase in liver, muscles and kidney of *Chana gachua. Indian J. Exp. Biol.*, 16: 1099-1102.

Danielli, J.F. 1952. Structural factors in cell permeability and secretion. In: Structural Aaspects of Cell Physiology. Symposia of the Society of Experimental Biology, Vol. 6, *Academic Press, Inc.*, New York, pp. 1-15.

Das, C.C., B.P. Kaufmann and H. Gay, 1964. Histone-protein transition in *Drosophilla malanogaster*: I. Changes during spermatogenesis. *Exp. Cell Res.*, 35(3): 507-514.

De Duve, C. 1959. Lysosomes, a new group of cytoplasmic particles. In: Subcellular particles. Hayasbi, T. (ed.), *Ronald press*, New York, pp.128-159.

Deb, D.C. and S. Chakravorty, 1981. Morphogenetic changes induced by additional corpora allata in ovary of the rice moth Corcyra cephalonica Stainton. *Indian J. Exp. Biol.*, 19(10): 922-925.

Deb, D.C. and S. Chakravorthy, 1982. Effect of precocene II, applied independently or subsequent to hydroprene treatment, on the morphogenesis of female reproductive organs of *Corcyra cephalonica. J. Insect Physiol.*, 28(8): 703-712.

Deb, D.C. and S. Chakravorty, 1985. Influence of additional corpora allata, juvenoids and antiallayotropin on the development and phenotypic changes of the rice moth *Corcyra cephalonica* (Stainton). *Insect Sci. Appl.*, 6(1): 105-110.

Dedos, S.G. and H. Fugo, 1999. Induction of dauer larvae by application of fenoxycarb early in the 5th instar of the silk worm, *Bombyx mori. J. Insect Physiol.*, 45(8): 769-775.

Delisle, J. and M. Cusson, 1999. Juvenile hormone biosynthesis, oocyte growth and vitellogenin accumulation in *Choristoneura fumiferana* and *C. rosaceana*: a comparative study. *J. Insect Physiol.*, 45(6): 515-523.

Denlinger, D.L. 1985. Hormonal control of diapause. In: *Comprehensive Insect Physiology, Biochemistry and Pharmacology*. Kerkut, G.A. and L.I. Gilbert (eds.), Vol. 8, *Pergamon Press,* Oxford, pp. 353412.

Devi, M.B., N.V. Devi, S.R. Devi and P.R. Singh, 2013. Biology and morphometric of rice moth, *Corcyra cephalonica. Ann. Plant Prot. Sci.*, 21(1): 87-89.

DeVries, D.H. and T.M. Brown 1977. Significance of the urogomphal abnormality as a criterion in the test method for susceptibility to JH mimics in *Tribolium confusum*. *J. Econ. Entomol.*, 70: 273-276.

Dhadialla, T.S. and G. Tzertzinis, 1997. Characterization and partial cloning of ecdysteroid receptor from a cotton boll weevil embryonic cell line. *Arch. Insect Biochem. Physiol.*, 35: 45-57.

Dhadialla, T.S., G.R. Carlson and D.P. Le, 1998. New insecticides with ecdysteroidal and juvenile hormone activity. *Ann. Rev. Entomol.*, 43: 545-569.

Dhadialla, T.S., A. Retnakaran and G. Smagghe, 2005. Insect growth and development-disrupting insecticides. In: Comprehensive Insect Molecular Science. Gilbert, L.I., I. Kostas and S.S. Gill (eds.), Vol. 6, *Pergamon; Elsevier,* Oxford, 55-115.

Dorn, S., M.L. Frishknecht, V. Martinez, R. Zurfluh and U. Fisher, 1981. A novel non-neurotoxic insecticide with a broad activity spectrum. *Z. Pflanzenk. Pflanzen.*, 88(5): 269-275.

Dubrovski, E.B., V.A. Dubrovskaya, A.L. Bilderback and E.M. Berger, 2000. The isolation of two juvenile hormone-inducible genes in *Drosophila melanogaster. Dev. Biol.*, 224(2): 486-495.

Dumser, J.B. and K.G. Davey, 1974. Endocrinological and other factors influencing testis development in *Rhodnius prolixus. Can. J. Zool.*, 52(8): 1011-1022.

Dunkel, F.V. and L.J. Sears, 1998. Fumigant properties of physical preparations from mountain big sagebrush artemisia tridentate Nutt. Ssp. *J. Stored Prod. Res.*, 34: 304-321.

Dyte, C.E. 1972. Resistance to synthetic juvenile hormone in a strain of the flour beetle, *Tribolium castaneum. Nature,* 238: 48-49.

Edwards, J.P. 1976. Age-related susceptibility of *Tribolium castaneum* (Herbst) to synthetic C_{18} juvenile hormone. *J. Stored Prod. Res.*, 12(2): 71-76.

Edwards, J.P. and J.J. Menn, 1981. The use of juvenoid in insect pest management. In: Insektizide · Bakterizide · Oomyceten-Fungizide / Biochemische und biologische Methoden · Naturstoffe / Insecticides · Bactericides · Oomycete Fungicides / Biochemical and Biological Methods · Natural Products. Wegler, R. (ed.), Vol. 6, *Springer*, Berlin Heidelberg, pp. 185-214.

Edwards, J.P. and J.E. Short, 1984. Evaluation of three compounds with insect juvenile hormone activity as grain protectants against insecticide-susceptible and resistant strains of *Sitophilus* species (Coleoptera: Curculionidae). *J. Stored Prod. Res.*, 20(1): 31-39.

Edwards, J.P. and L. Abraham, 1985. Laboratory evaluation of two insect juvenile hormone analogues against *Alphitobius diaperinus* (Panzer) (Coleoptera: Tenebrionidae). *J. Stored Prod. Res.*, 21(4): 189-194.

Edwards, J.P., J.E. Short and L. Abraham, 1987. New approaches to the protection of stored products using an insect juvenile hormone analogue. British Crop Protection Council, Monograph no. 37, *Stored Prod. Pest Control,* pp. 197-205.

Edwards, J.E., J.E. Short and L. Abraham, 1991. Large-scale evaluation of the insect juvenile hormone analogue fenoxycarb as a long-term protectent of stored wheat. *J. Stored. Prod. Res.*, 27(1): 31-39.

Eisa, A.A. 1992. Chemical control of Angoumois grain moths, *Sitotroga cerealella* Olivier (Lepidopter: Gelechiidae). *Emir. J. Agric. Sci.*, 4: 80- 90.

El-barky, N.M., O.M. El-Monairy, R.F.A. Bakr and N.M.B. El-shourbagy, 2012. Biological and behavioural effects of pyriproxyfen on pheromone production and perception of *Tribolium castaneum* (Coleoptera: Tenebrionidae). *Egypt. Acad. J. Biolog. Sci.*, 4(1): 11-22.

El-Bassal, M.M. 1993. Combined effects of insect growth regulators and gamma radiation on *Culex pipiens*. Ph.D. Thesis, Faculty of Science, Cairo Univercity, Giza, Egypt.

El-Sayed, F.M.A., 1984. Effect of the synthetic insect growth regulator methoprene on larval development and reproduction of two species of stored product insects. *Bull. Soc. Entomol.*, 65: 215-221.

El-Sayed, F.M.A. 1988. Biological activity of the synthetic insect growth regulator hydro-prene against stored product coleopterand. *Agric. Res. Rev.*, 66: 63-69.

El-Sheikh, T.A. 2002. Effects of application of selected insect growth regulators and plant extracts on some physiological aspects of the black cutworm, *Agrotis ipsilon* (Huf). Ph.D. Thesis, Faculty of Science, Ain shams University. Cairo, Egypt.

El-Sherif, L. S. 1995. Effect of juvenile hormone analogue, pyriproxyfen on the main metabolites in the haemolymph of last instar nymph of *Schistocerca gregaria* (Orthoptera: Acrididae). *J. Egypt. Ger. Soc. Zool.*, 16(E): 125-39.

Elvira, S., T. Williams and P. Caballero, 2010. Juvenile hormone analog technology: effects on larval cannibalism and the production of *Spodoptera exigua* (Lepidoptera: Noctuidae) nucleopolyhedrovirus. *J. Econ. Entomol.*, 103(3): 577-582.

Engles, W. and W. Drescher, 1964. Einbau von H^3-D- glucose wahrend der oogenase bei *Apis mellifica* (L.). *Experientia*, 20(8): 445-447.

Engles, W. 1966. Der zeitliche ablauf von protein- und kohlenhydratsynthesen in der oogenase bei *Apis mellifica* L. *Zool. Anz. Suppl.*, **29**: 243-251.

Engles, W. and K. Bier, 1967. Zur glycogenspeicherung wahrend oogenase und ihrer vorzeitiger auslosung durch blockiering der RNS-versorgung (untersuchungen an *Musca domesticus* L.). *Wilhelm Roux Arch. Entwick. -Mech. Org.*, 158: 64-88.

Etebari, K., A.R. Bizhannia, R. Sorati and L. Matindoost, 2007. Biochemical changes in haemolymph of silkworm larvae due to pyriproxyfen residue. *Pestic. Biochem. Physiol.*, 88: 14-19.

Extension Toxicology Network. 1993. Wave page: http://ace.orst.edu/cgi-bin/mfs/01/pips/methopr.

Fajardo, V. and B. Morallo-Rejesus, 1980. Effect of ten insect growth regulators on *Plodia interpunctella* Hubn. and *Corcyra cephalonica* Staint. *Philippine Entomol.*, 4(3): 169-175.

FAO/WB Report, 2011. Stemming post-harvest waste crucil to African food security: *Grain losses in sub-Saharan Africa could total $4 billion*, Press release no: 2011/522/AFR. Online url: http://go.worldbank.org/T35HKXAG70

Feng, Q.L, T.R. Ladd, B.L. Tomkins, M. Sundaram, S.S. Sohi, A. Retnakaran, K.G. Davey and S.R. Palli, 1999. Spruce budworm (*Choristoneura fumiferana*) juvenile hormone esterase: hormonal regulation, developmental expression and cDNA cloning. *Mol. Cell. Endocrinol.*, 148(1-2): 95-108.

Ferenz, H.J. 1963. Juvenile hormone induced sexual maturity in males of *Pterostichus nigrita* F. *Experienti*, 29: 356-366.

Finney, D.J. 1952. Probit Analysis.2nd Edn. *Cambridge Univ. Press, London*, pp 1-318.

Fitzpatrick, G.E. and R.V. Dowell, 1981. Survival and emergence of citrus blackfly parasitoids after exposure to insecticides. *Environ. Entomol.*, 10(5): 728-731.

Fox, P. 1990. Insect Growth regulators. *P.J.B. Publication Limited.* Richmond. U.K., pp. v + 108.

Friedman, S. 1970. Chemical Zoology. Florkin, M. and B.T. Scheer (eds.), Vol. 5, Part A, *Academic press*, New York, pp. 167-197.

Fukuda, S. 1944. The hormonal mechanism of larval molting and metamorphosismin the silkworm. *J. Fac. Sci. Tokyo Univ. Sect. IV*. 6: 477-532.

Gadenne, C., S. Grenier, G. Plantevin and B. Mauchamp, 1990. Effects of a juvenile hormone mimetic, fenoxycarb, on post-embryonic development of the European corn borer, *Ostrinia nubilalis* Hbn. *Experientia*, 46(7): 744-747.

Ghasemi, A., J. Sendi and M. Ghadamyari, 2010. Physiological and biochemical effect of pyriproxyfen on Indian meal moth *Plodia interpunctella* (Hubner) (Lepidoptera: Pyralidae). *J. Plant Prot. Res.*, 50 (4): 416-422.

Gilbert, L.I., N.A. Granger and R.M. Roe, 2000. The juvenile hormones: historical facts and speculations on future research directions. *Insect Biochem. Mol. Biol.,* 30(8-9): 617-644.

Gillott, C. and S.B. Gaines, 1992. Endocrine regulation of male accessory gland development and activity. *Can. Entomol.,* 124: 871-886.

Gillott, C. 1996. Male insect accessory glands: functions and control of secretory activity. *Invertebr. Repr. Dev.,* 30: 199-205.

Goldsworthy, G.J. and G. Gade, 1983. The chemistry of hypertrehalosemic factors. In: Endocrinology of Insects. Downer, R.G.H. and H. Laufer, (eds.), Vol. 1, *Alan R. Liss,* New York, pp. 109-119.

Goodman, W.G. 1990. A simplified method for synthesizing juvenile hormone-protein conjugates. *J. Lipid Res.,* 31(2), 354-357.

Goodman, W.G. and N. Granger, 2005.The juvenile hormones. In: Comprehensive Molecular *Insect Science.* Gilbert L.I., K. Iatrou and S.S. Gill (eds.), Vol. 3, *Elsevier Pergamon,* Oxford, pp. 319-408.

Grain Depot Fund Prospectus, 2011. Real estate investment portfolio for grain storage in India. International impact investing challenge, where finance meets impact, Available at- sustainableinvestingchallenge.org/b6.pdf

Grenier, S. and A.-M. Grenier, 1993. Fenoxycarb, a fairly new insect growth regulator: review of its effects on insects. *Ann. Appl. Biol.,* 122(2): 369-403.

Gupta, A.P. and J.M. Mkhize, 1983. Developmental and morphogenetic effects of two insect growth regulators (Hydroprene and R-20458) on the female rice weevil, *Sitophilus oryzae* (L.) (Coleoptera : Curculionidae). *App. Ent. Zool.,* 18(1): 11-21.

Hami, M., F. Taibi and N. Soltani-Mazouni, 2004. Effects of flucyclox-uron, a chitin synthesis inhibitor, on reproductive events and thickness of chorion in mealworms. *Commun. Agric. Appl. Biol. Sci.,* 69(3): 249-255.

Hartfelder, K. 2000. Insect juvenile hormone: from "status quo" to high society. *Braz. J. Med. Biol. Res.,* 33(2): 157-177.

Hayward, D.C., T.S. Dhadialla, S. Zhou, M.J. Kuiper, E.E. Ball, G.R. Wyatt and V.K. Walker, 2003. Ligand specificity and developmental expression of RXR and ecdysone receptor in the migratory locust. *J. Insect Physiol.,* 49(12): 1135-1144.

Henrick, C.A., G.B. Staal and J.B. Siddall, 1973. Alkyl 3, 7, 11-trimethyl-2, 4-dodecadienoates, a new class of potent insect growth regulators with juvenile hormone activity. *J. Agric. Food Chem.,* 21(3): 354-359.

Herford, G.V.B. 1933. The more important insect pests of cacao, tobacco and dried fruit. *Bull. Imp. Inst.,* 31(1): 39-55.

Hicks, B.J. and R. Gordon, 1992. Effects of the juvenile hormone analog fenoxycarb on various developmental stages of the eastern spruce budworm, *Choristoneura fumiferana* (Clemens) (Lepidoptera: Tortricidae). *Can. Entomol.,* 124(1): 117-123.

Hill, D.S. 1990. Pests of stored products and their control. *Belhaven Press,* London, pp. 8-55.

Hinton, H.E. 1945. A Monograph of the beetles associated with stored products. Vol. I *Trustees of the British Museum (Natural History)*, London, p 443.

Hiruma, K. 2003. Juvenile hormone action in insect development. In: Encyclopedia of Hormones and Related Cell Regulators, Henry, H.L. and A.W. Norman (Eds.), vol. 2, *Academic Press*, San Diego, pp. 528-535.

Hoehn, H. and T. Wildbolz, 1992. Side effects of fenoxycarb and diflubenzuron on secondary pests in apple orchards. *Acta. Phytopathol. Entomol. Hungarica*, 27: 281-287.

Hoffmann, K.H. and M.W. Lorenz, 1997. The role of ecdysteroids and juvenile hormones in insect reproduction. *Trends Comp. Biochem. Physiol.*, 3: 1-8.

Hollander, V.P. 1971. Acid phosphatase. In: The Enzymes. Boyer, P.D. (ed.), Vol. 4, *Academic Press*, New York, pp. 449-498.

Hong, B.T.K. 1975. Effects of a synthetic juvenile hormone and some analogues on *Ephestia* spp. (Lepidoptera: Phycitidae). *Ann. Appl. Biol.*, 80(2): 137-145.

Hoppe, T. 1974. Effects of a juvenile hormone analogue on Mediterranean flour moth in stored grains. *J. Econ. Entomol.*, 67: 789.

Hoppe, T. 1976. Microplot trial with an epoxyphenylether (insect growth regulator) against several pests of stored wheat grain. *J. Stored Prod. Res.*, 12(3): 205-209.

Hoppe, T. 1981. Testing of methoprene in resistant strains of *Tribolium castaneum* (Herbst.) (Col., Tenebrionidae). *Z. Angew. Entomol.*, 91: 241-251.

Horowitz, A.R. and I. Ishaaya, 1994. Managing resistance in insect growth regulator in the sweet potato white fly (Homoptera: Aleyrodidae). *J. Econ. Entomol.*, 87(4): 866-871.

Ibrahim, A.M., M.G. Higazi and E.S. Demian, 1974. Histochemical localization of alkaline phosphatase activity in the alimantry tract of the snail, *Marisa cornuarietis* (L.). *Bull. Zool. Soc. Egypt.*, 26: 94-105.

Ishaaya, I. and S. Yablonski, 1976. Induction of prolonged larval feeding stage by hormone analogues in *Tribolium castaneum. Phytoparasitica*, 4(1): 9-18.

Ishaaya, I. 1982. Biological and biochemical aspects of the disruption of adult formation in *Tribolium* by a novel bis (thiocarbamate) R-31026 and the juvenoid R-20458. *Pestic. Sci.*, 13(2): 204-210.

Ishaaya, I., S. Yablonski and K.R.S. Ascher, 1987. Toxicological and biochemical aspects of novel acylureas on resistant and susceptible strains of *Tribolium castaneum. Proc. 4th Int. Work Conf. Stored-Product Prot.*, Donahaye E. and S. Navarro, (Eds.), *Tel Aviv*, Israeel, pp. 613-622.

Ishaaya, I. and A.R. Horwitz, 1998. Insecticides with novel modes of action: an overview. In: Insecticides with Novel Modes of Action: Mechanisms and Application, Ishaaya, I. and D. Degheele (eds.), *Springer*, Berlin Heidelberg, pp. 1-24.

Iyengar, A.R. and J.G. Kunkel, 1955. Follicle cell calmodulin in *Blatella gexmanica*: transcript accumulation during vitellogenesis is regulated or juvenile hormone. *Dev. Biol.*, 170: 314-320.

Jagadish, P.S., P. Nirmala, M.A. Rashmi, J.N. Hedge and N. Nangia, 2009. Biology of rice moth, *Corcyra cephalonica* Stainton on foxtail millet. *Karnataka J. Agric. Sci.*, 22(3): 674-675.

Jenson, E.A., F.H. Arthur and J.R. Nechols, 2010a. Efficacy of an esfenvalerate plus methoprene aerosol for the control of eggs and fifth instars of *Plodia interpunctella* (Lepidoptera: Pyralidae). *Insect Sci.*, 17: 21-28.

Jenson, E.A., F.H. Arthur and J.R. Nechols, 2010b. Methoprene and synergized pyrethrins as aerosol treatments to control *Plodia interpunctella* (Hübner), the Indian meal moth (Lepidoptera: Pyralidae). *J. Stored Prod. Res.*, 46(2): 103-110.

Jones, G., T. Hanzlik, B.D. Hammock, D.A. Schooley, C.A. Miller, L.W. Tsai and F.C. Baker, 1990. The juvenile hormone titre during the penultimate and ultimate larval stadia of *Trichoplusia ni*. *J. Insect Physiol.*, 36(2): 77-83.

Jones, G. and P.A. Sharp, 1997. Ultraspiracle: an invertebrate nuclear receptor for juvenile hormones. *Proc. Natl. Acad. Sci. U.S.A.*, 94: 13499-13503.

Jones, G., M. Wozniak, Y. Chu, S. Dhar and D. Jones, 2001. Juvenile hormone III-dependent conformational changes of the receptor ultraspiracle. *Insect Biochem. Mol. Biol.*, 32(1): 33-49.

Kaakeh, W., B.L. Reid, T.J. Bohnert and G.W. Bennett, 1997. Toxicity of imidacloprid in the german cockroach (Dictyoptera: Blattellidae), and the synergism between imidacloprid and Metarhi- zium anisopliae (Imperfect Fungi: Hyphomycetes). *J. Econ. Entomol.*, 90: 473-482.

Kamimura, M. 1995. Effect of a juvenile hormone analogue, fenoxycarb, on larval growth of the silk worm, *Bombyx mori* (Lepidoptera: Bombycidae). *Appl. Entomol. Zool.*, 30: 487-489.

Kamimura, M. and M. Kiuchi, 1998. Effects of a juvenile hormone analog, fenoxycarb, on 5th stdium of the larvae of the silkworm, *Bombyx mori* (Lepidoptera: Bombycidae). *Appl. Entomol. Zool.*, 33: 333-338.

Kamimura, M. and M. Kiuchi, 2002. Applying fenoxycarb at the penultimate instar triggers an additional ecdysteroid surge and induces perfect extra larval molting in silk worm. *Gen. Comp. Endocrinol.*, 128(3): 231-237.

Kanost, M.R., J.K. Kawooya, J.H. Law, R.O. Ryan, M.C. van Heusden and R. Ziegler, 1990. Insects haemolymph proteins. *Adv. Insect Physiol.*, 22: 299-396.

Kavallieratos, N.G., C.G. Athanassiou, B.J. Vayias and Z. Tomanovic, 2012. Efficacy of insect growth regulators as grain protectants against two stored-product pests in wheat and maize. *J. Food Prot.*, 75(5): 942-950.

Kawatra, V.K. 1976. Studies on the effect of tepa on certain physiological processes associated with the reproduction in the female tropical house mosquito *Culex pipiens fatigans*, Ph. D. Thesis, University of Delhi, India.

Kawatra, V.K. 1981. Effects of tepa on alkaline phosphatase activity during ovarian development of the tropical house mosquito *Culex pipiens fatigans* Wiedemann. *Indian J. Exp. Biol.*, 19(3): 212-215.

Keeley, L.L. 1978. Endocrine regulation of fat body development and function. *Ann. Rev. Entomol.*, 23: 329-352.

Keeley, L.L. and S.R. McKercher, 1985. Endocrine regulations of ovarian maturation in the cockroach *Blaberus discoidalis. Comp. Biochem. Physiol. A: Comp. Physiol.*, 80(1): 115-121.

Kellouche, A. and N. Soltani, 2006. Impact of hexaflumuron, a chitin synthesis inhibitor, on growth, development and repro-ductive performance of the progeny in *Callosobruchus mac-ulatus* after adult treatments. *Afr. J. Agric. Res.*, 1(3): 57-64.

Khebbeb, M.E.H., R. Gaouaoui and F. Bendjeddou, 2008. Tebufenozide effects on the reproductive potentials of the Mediterranean flour moth, *Ephestia kuehniella. African J. Biotech.*, 7(8): 1166-1170.

Kidd, H. and D.R. James, 1991 (eds.). The Agrochemicals Handbook, Third Edition. *Royal Society of Chemistry Information Services*, Cambridge, UK, pp. 3-11.

Kilby, B.A. 1963. The biochemistry of the insect fat body. In: Advances in Insect Physiology. Beament, J.W.L., J.E. Treherne and V.B. Wigglesworth (eds.), Vol. 1, *Academic Press*, London and New York, pp. 111-174.

Kim, Y., D. Kima and J. Lee, 2000. Disturbance of adult eclosion by fenoxycarb, a juvenile hormone mimic, in the beet armyworm, *Spodoptera exigua. J. Asia-Pacific Entomol.*, 3(2): 103-111.

Koelle, G.B. 1975. Neurohumoral transmission and the autonomic nervous system. Anticholinesterase agents In: The Pharmacological basis of Therapeutics. Goodman, L.S. and A. Gilman (eds.), fifth edition, *Macmillan*, New York, pp. 404-466.

Koeppe, J.K., K. Hobson and S.E. Wellman, 1980. Juvenile hormone regulation of structural changes and DNA synthesis in the follicular epithelium of *Leucophaea maderae. J. Insect Physiol.*, 26(4): 229-240.

Koeppe, J.K., M. Fuchs, T.T. Chen, L.M. Hunt, G.E. Kovalick and T. Briers, 1985. The role of juvenile hormone in reproduction. In: Comprehensive Insect Physiology Biochemistry and Pharmacology. Kerkut, L.I. and G.A. Gilbert, (ed.), Vol. 8, *Pergamon Press*, Oxford, pp. 165-203.

Kopec, S. 1922. Studies on the necessity of the brain for the inception of insect metamorphosis. *Biol. Bull.*, 42(6): 323-342.

Kostyukovsky, M., B. Chen, S. Atsmi and E. Shaaya, 2000. Biological activity of two juvenoids and two ecdysteroids against three stored product insects. *Insect Biochem. Mol. Biol.*, 30 (8-9): 891-897.

Kramer, K.J., H.E. McGregor and K. Mori. 1979. Susceptibility of stored-product insects to pyridyl ether analogues of juvenile hormone. *J. Agric. Food Chem.* 27(6): 1215-1217.

Kramer, K.J. and G.B. Staal, 1981. In vitro studies on the mechanism of action of anti-juvenile hormone agents in larvae of *Manduca sexta* (L.). In: Juvenile Hormone Biochemistry-Action, Agonism and Antagonism. Pratt, G.E. and G.T. Brooks (eds.), *Elsevier/North-Holland Biomedical Press,* Amsterdam, pp: 425-437.

Kramer, K.J., R.W. Beeman and L.H. Hendricks, 1981. Activity of RO 13-5223 and RO 13-7744 against stored product insects. *J. Econ. Entomol.,* **74**: 678-680.

Kramer, K.J., L.H. Hendricks, J.H. Wojciak and J. Fyler, 1985. Evaluation of fenoxycarb, *Bacillus thuringiensis,* and malathion as grain protectants in small bins. *J. Econ. Entomol.,* 78(3): 632-636.

Krishna, S.S. and A.S. Narain, 1976. Ovipositional programming in the rice-moth, *Corcyra cephalonica* (Staint.) (Lepidoptera: Galleridae) in relation to certain extrinsic and intrinsic caues. *Proc. Indian Natl. Sci. Acad.,* B 42: 325-332.

Kunkel, J.G. and J.H. Nordin, 1985. Yolk protein. In: Com-parative Insect Biochemistry, Physiology, and Pharmacol-ogy. Kerkut, G.A. and L.I. Gilbert, (eds.), Vol.1, *Pergamon,* Oxford, pp. 83-112.

Lam, K.C. and S.P. Mistills, 1973. Role of intestinal alkaline phosphatase in fat transport. *Aust. J. Exp. Biol. Med. Sci.,* 51(3): 411-416.

Lang, C.A., H.Y. Lau and D.J. Jefferson, 1965. Protein and nucleic acid changes during growth and aging in the mosquito. *Biochem. J.,* 95: 372-377.

Lee, J.J., J.C. Shim and K.R. Lee, 1981. The effect of insecticides on haemolymph in the German cockroach, *Blattela germanica* L. *Korean J. Entomol.,* 11: (1) 25-32.

Lee, K.-Y., M.E. Chamberlin and F.M. Horodyski, 2002. Biological activity of *Manduca sexta* allalotropin-like peptides, predicted products of tissue-specific and developmentally regulated alternatively spliced mRNAs. *Peptides,* 23: 1933-1941.

Leighton, T.E., D. Marks and F. Leighton, 1981. Pesticides: insecticides and fungicides are chitin synthesis inhibitors. *Science,* 213(4510): 905-907.

Letellier, C., E. Haubruge and C. Gasper, 1995. Biological activity of *fenoxycarb* against *Sitophilus zeamais* Motsch. (Coleoptera: Curculionidae). *J. Stored Prod. Res.,* 31(1): 37-42.

Limonta, L., D.P. Locatelli, T. Broglia and J. Baumgartner, 2009. Cohort development models for integrated *Corcyra cephalonica* (Stainton) population management. *Boll. Zool. Agr. Bachic. Ser.* II., 41(3): 215-226.

Liu, T.P. 1973. The effect of allatectomy on the blood trehalose in the female house fly, *Musca domestica* L, *Comp. Biochem. Physiol. A.,* 46(1): 109-113.

Liu, T.P. 1974. The effect of allatectomy on glycogen metabolism in the fat-body of the female house fly, *Musca domestica* L. *Comp. Biochem. Physiol. B.,* 47(1): 79-86.

Liu, T.-X. and T.-Y. Chen, 2001. Effects of the insect growth regulator fenoxycarb on immature *Chrysoperla rufilabris* (Neuroptera: Chrysopidae). *Flo. Entomol.,* 84(4): 628-633.

Loni, S., M. Moarefi, H. Farazmand and E. Karami, 2011. The effect of regulating compounds on the growth of *Tribolium confusum* du Val (Coleoptera: Tenebrionidae). *Mun. Ent. Zool.*, 6(1): 377-385.

Loschiavo, S.R. 1975. Tests of four synthetic insect growth regulators with juvenile hormone activity against seven species of stored products insects. *Manit. Entomol.*, 9: 43-52.

Loschiavo, S.R. 1976. Effects of the synthetic insect growth regulators methoprene and hydroprene on survival, development or reproduction of six species of stored-products insects. *J. Econ. Entomol.*, 69(3):395-399.

Lowry, O.H., N.J. Rosenbrough, A.L. Farr and R.J. Randall, 1951. Protein measurement with the Folin phenol reagent. *J. Biol. Chem.*, 193: 265-275.

Ludwig, S.W. and R.D. Oetting, 2001. Evaluation of medium treatments for management of *Frankliniella occidentalis* (Thripidae: Thysanoptera) and *Bradysia coprophila* (Diptera: Sciaridae). *Pest Manag. Sci.*, 57(12): 1114-1118. doi: 10.1002/ps.404

Mamatha, D.M., V.K. Kanji, H.H. Cohly and M.R. Rao, 2008. Juvenile hormone analogues, methoprene and fenoxycarb dose-dependently enhance certain enzyme activities in the Silkworm *Bombyx mori* (L). *Int. J. Environ. Res. Public Health*, 5(2): 120-124.

Mandal, S. 1982. Effect of juvenoid and allatectomy on the biochemical components of gonads in *Gryllotalpa gryllotalpa* (Grllotalpidae: Orthoptera: Insecta). *Proc. Indian Natl. Sci. Acad.*, B48(4): 486-492.

Marx, J.L. 1977. Chitin Synthesis Inhibitors: New Class of Insecticides. *Science*, 197: 1170-1172.

Masner, P., S. Dorn, W. Vogel, M. Kalin, O. Graf and E. Gunthart, 1980. Types of response of insects to a new IGR and to proven standard insecticides. In: Regulation of Insect Development and Behavior. Kloza, M. (ed.). *Technical University Wroclaw Press*, Poland. pp: 809-818.

Masner, P., K. Salama and V. Landa, 1968. Natural and synthetic materials with insect hormone activity. IV. Specific female sterility effects produced by a juvenile hormone analogue. *J. Embryol. Exp. Morph.*, 20(1): 25-31.

Masner, P., M. Angst and S. Dorn, 1987. Fenoxycarb, an insect growth regulator with juvenile hormone activity: a candidate for *Heliothis virescens* (F.) control on cotton. *Pestic. Sci.*, 18(2): 89-94.

Matolcsy, G., M. Nadasy and V. Andriska, 1988. Pesticide Chemistry: Studies in Environmental Science. *Elsevier*, New York, pp. 808.

Mauchamp, B., C. Malosse and P. Saroglia, 1989. Biological effects and metabolism of fenoxycarb after treatment of the fourth and the fifth instars of tobacco budworm, *Heliothis virescense* F. *Pest Sci.*, 26(3): 283-301.

Mazlan, N. and J. Mumford, 2005. Insecticide use in cabbage pest management in the Cameron Highlands, Malaysia. *Crop Prot.*, 24(1): 31-39.

Mbata, G.N. and T.W. Phillips, 2001. Effects of temperature and exposure time on mortality of stored-product insects exposed to low pressure. *J. Econ. Entomol.*, 94(5): 1302-1307.

McDonald, J. 1995. Terrestrial field dissipation of 14C-fenoxycarb 25WP on bareground soil in California, *Ciba-Geigy Corporation Data Package Report* No. 163859-N. DPR No. 51317-085.

McGregor, H.E. and K.J. Kramer, 1975. Activity of insect growth regulators, hydroprene and methoprene, on wheat and corn against several stored-grain insects. *J. Econ. Entomol.*, 68(6): 668-670.

Metcalf, R.L. 1968. Methods of estimating effects. In: Research in Pesticides. C.O. Chichester, ed. Academic Press, New York. pp 17-29.

Metcalf, R.L. 1971. The chemistry and biology of pesticides. In: Pesticides in the environment. R. White-Stevens (Ed.), Vol. I, *Marcel Dekker Inc.* New Yark, pp. 1-144.

Metwally, M.M. and V. Landa, 1972. Sterilization of the khapra beetle, *Trogoderma granarium* Everts, with juvenile hormone analogues. *Z. Angew. Entomol.*, 72(1-4): 97-109.

Mian, L.S. and M.S. Mulla, 1982a. Biological activity of IGRs against four stored-product coleopterans. *J. Econ. Entomol.*, 75(1): 80-85.

Mian, L.S. and M.S. Mulla, 1982b. Residual activity of insect growth regulators against stored-product beetles in grain commodities. *J. Econ. Entomol.*, 75(4): 599-603.

Miao, Y. 2002. Studies on the activity of the alkaline phosphatase in the midgut of infected silkworm, *Bombyx mori* L., *J. Appl. Entomol.*, 126(2-3): 138-142.

Miranda, J.E., S.A. deBortoli and R. Takahashi, 2002. Development and silk production by silkworm larvae after topical application of methprene. *Sci. Agric.*, 59(3): 585-588.

Mirth, C.K., H.Y. Tang, S.C. Makohon-Moore, S. Salhadar, R.H. Gokhale, R.D. Warner, T. Koyama, L.M. Riddiford and A.W. Shingleton, 2014. Juvenile hormone regulates body size and perturbs insulin-signaling in *Drosophila. Proc. Natl. Acad. Sci.* U.S.A., 111(19): 7018-7023.

Mishra, S.N. and S.S. Krishna, 1980. Oviposition and egg fertility in *Corcyra cephalonica* Staint. (Lepidoptera: Galleridae) affected by programmed manipulations in larval ecology during rearing or in adult light regimes. *Z. Angew. Zool.*, 67(1): 101-105.

Mitilin, N., G. Wiygul and J.W. Haynes, 1977. Inhibition of DNA synthesis in boll weevils, (*Anthnomus grandis* Boheman) sterilized by dimilin. *Pestic. Biochem. Physiol.*, 7(6): 559-563.

Miura, T., C.H. Schaefer and R.J. Stewart, 1986. Impact of carbamate insect growth regulators on the selected aquatic organisms; a preliminary study. *Proc. Ann. Conf. California Mosquito Vector Control Assoc.*, 54: 36-38.

Miura, T. and R.M. Takahashi, 1987. Impact of fenoxycarb, a carbamate insect growth regulator, on some aquatic invertebrates abundant in mosquito breeding habitats. *J. Amer. Mosquito Control Assoc.*, 3: 476-480.

Miyamoto, J., M. Hirano, Y. Takimoto and M. Hatakoshi, 1993. Insect growth regulators for pest control, with emphasis on juvenile hormone analogs: Present Status and Future Prospects. In: Pest Control with Enhanced Environmental Safety. Duke, S.O., J.J. Menn and J.R. Plimmer (eds.), Vol. 524, *ACS Symposium Series*, pp. 144-168.

Mkhize, J.N. and A.P. Gupta, 1982. Effects of two insect growth regulators (Hydroprene and R-20458) on the follicular epithelium and the oocytes of the rice weevil, *Sitophilus oryzae* (L.) (Coleoptera: Curculionidae). *New Entomol. Soc.*, 90(3): 213-218.

Mkhize, J.N. 1988. Synthetic juvenile analogues against four species of stored products beetles. *Intr. J. Trop. Insect. Sci.*, 9(2): 275-278.

Moczon, T. 1976. Inhibitory effect of pesticides on phosphatase, glucosidase and acetylcholinesterase activity in miracidium of *Fasciola hepatica*. Histochemical data. *Bull. Acad. Pol. Sci. Ser. Sci. Biol.*, 24(5): 289-292.

Mohandas, S., F.H. Arthur, K.Y. Zhu and J.E. Throne, 2006a. Hydroprene prolongs developmental time and increases mortality of Indianmeal moth (Lepidoptera: Pyralidae) eggs. *J. Econ. Entomol.*, 99(3):1007-1016.

Mohandas, S., F.H. Arthur, K.Y. Zhu and J.E. Throne, 2006b. Hydroprene prolongs development time and increases mortality in wanderingphase Indianmeal Moth (Lepidoptera: Pyralidae) larvae. *J. Econ. Entomol.*, 99(4): 1509-1519.

Mojaver, M. and A.R Bandani., 2010. Effects of the insect growth regulator pyriproxyfen on immature stages of Sunn Pest, *Eurygaster integriceps* Puton (Heteroptera: Scutelleridae). *Mun. Entomol. Zool.*, 5 (1): 187-197.

Monconduit, H. and B. Mauchamp, 1999. Fenoxycarb, a potent contaminant of the silk worm *Bombyx mori* L., does not influence its juvenile hormone titer. *Arch. Insect Biochem. Physiol.*, 40(3): 141-149.

Mondal, K.A.M.S.H. and G.R. Port, 1995. Effects of cyromazine on larval growth and adult population of susceptible and malathion-resistant striains of *Tribolium castaneum* Herbst. *J. Bio. Sci.*, 3:1-10.

Mondal, K.A.M.S.H. and S. Parween, 2000. Insect growth regulators and their potential in the management of stored-product insect pests. *Integ. Pest Manag. Rev.*, 5(4): 255-295.

Moreno, J., N. Hawlitzky and R. Jimenez, 1992. Effect of the juvenile hormone analog fenoxycarb on the last larval instar of *Ephestia kuehniella* Zell. (Lep., Pyralidae). *J. Appl. Entomol.*, 114(1-5): 118-123.

Morohoshi, S. and K. Kiguchi, 1969. Effect of corpus allatum hormone on lipid metabolism in *Bombyx mori*. *Proc. Japan Acad.*, 45(4): 323-327.

Muckenthaler, F.A. 1964. Autoradiographic study of nucleic acid synthesis during spermatogenesis in the grasshopper, *Malanoplus differentialis. Exp. Cell Res.*, 35: 531-547.

Mulder, R. and M.J. Gijswijt, 1973. The laboratory evaluation of two promising new insecticides which interfere with cuticle deposition. *Pestic. Sci.* 4(5): 737-745.

Mulye, H. and R. Gordon, 1989. Effects of selected juvenile hormone analogue hormone analogues on sixth-instar larvae of the eastern spruce budworm, *Choristoneura fumiferana* Clemens (Lepidoptera: Tortricidae). *Can. Entomol.* 121(12): 1111-1116.

Munro, J.W. and W.S. Thompson, 1929. Report on insect infestation of Cacao. London: H.M. Stationery Office, 24.

Murali Baskaran, R.K. and R. Janarthanan, 1991. Effect of insect growth regulators on longevity of three species of stored products pests. *Bull. Grain Technol.*, 29(1): 35-37.

Nagpal, M. and A. Kumar, 2012. Grain losses in India and government policies. *Qual. Assur. Saf. Crop.*, 4(3): 143. doi: 10.1111/j.1757-837X.2012.00150.x

Naqvi, S.N.H., S. Shafi and N. Zia, 1989. Histochemical localization of proteins and alkaline phosphatases in diflubenzuron treated insects. *Pakistan J. Sci. Ind. Res.*, 32:190-193.

National Research Council, Commission on International Relations, Board of Science and Technology for International Development, 1978. Postharvest Food Losses in Developing Countries. Staff summary report of an international working group meeting, Washington, D.C., National Academy of Sciences. 1e

Nicholas, A.H., R.N. Spooner-Hart and R.A. Vickers, 2005. Abundance and natural control of the woolly aphid *Eriosoma lanigerum* in an Australian apple orchard IPM program, *BioControl*, 50(2): 271-291.

Nickle, D.A. 1979. Insect growth regulators: New protectants against the almond moth in stored inshell peanuts. *J. Econ. Entomol.*, 72: 816-819.

Niimi, S. and S. Sakurai, 1997. Developmental changes in juvenile hormone acid titers in the hemolymph and in-vitro juvenile hormone synthesis by corpora allata of the silkworm, *Bombyx mori. J. Insect Physiol.*, 43(9): 875-884.

Nijhout, H.F. 1994. Insect Hormones. *Princeton University Press*, Princeton, NJ. pp: 267.

Nijhout, H.F. 1999. Control mechanisms of polyphenic development in insects: In polyphenicd evelopment, environmentalfa ctorsa lters ome aspects of development in an orderly and predictable way. *Bio Sci.*, 49(3): 181-192.

Nino, E.L., C.E. Sorenson, S.P. Washburn and D.W. Watson, 2009. Ef-fects of the insect growth regulator, methoprene, on *On-thophagus taurus* (Coleoptera: Scarabaeidae). *Environ. Entomol.*, 38(2): 493-498.

Nordlander, G. 1975. Tests with altozar and altosid against two insects pests of stored grain. *Vaxtskyddsnotiser*, 39(6): 136–9.

Noyes, W.M. 1930. Moth pests in cocoa and confectionery. *Bull. Entomol. Res.*, 21(1): 77-121.

O'Brien, R.D. 1976. Acetylcholinesterase and its inhibition. In: Insecticides Biochemistry and Physiology. Wilkinson, C.F. (ed.), *Plenum Press*, Springer US. pp. 271-296,

Oberlander, H. and D.L. Silhacek, 1976. Action of juvenile hormone on imaginal discs of Indian meal moth. In: The Juvenile Hormones. Gilbert L.I. (ed.), *Springer US*, New York, pp. 220-233.

Oberlander, H., D. Nickle, D.L. Silhacek and D.W. Hagstrum, 1978. Advances in insect growth regulators research with grain insects. *Symposium on the Prevention and Control of Insects in Stored Foods Products. Manhattan, Kansas*. p. 247-263.

Oberlander, H., D.L. Silhacek, E. Shaaya and I. Ishaaya, 1997. Current status and future perspectives of the use of insect growth regulators for the control of stored product insects. *J. Stored Prod. Res.*, 33(1): 1-6.

Odhiambo, T.R. 1966. The metabolic effects of corpus allatum hormone in the male desert locust. I. Lipid metabolism. *J. Exp. Biol.*, 45(1): 45-50.

Olivieri, G. and A. Olivieri, 1965. Autoradiographic study of nucleic acid synthesis during spermatogenesis in *Drosophila melanogaster*. *Mutat. Res./Fund. Mole. M.* 2(4): 366-380.

Palli, S.R. and A. Retnakaran, 2000. Ecdysteroid and juvenile hormone receptors: properties and importance in developing novel insecticides. In: Biochemical Sites of Insecticide Action and Resistance. Ishaaya, I. (ed.), *Springer*, Berlin Heidelberg, pp. 107-132.

Pallos, F.M., J.J. Menn, P.E. Letchworth and J.B. Miaullis, 1971. Synthetic mimics of insect juvenile hormone. *Nature*, 232: 486-487. doi:10.1038/232486a0

Palm, N.B. 1948. Normal and pathological histology of the ovaries in *Bombus* Latr. (Hymenopt.). *Opuscula Entomol. Supplement.*, 7: 1-110.

Paoletti, M.G. and D. Pimental, 2000. Environmental risks of pesti-cides versus genetic engineering for agricultural pest con-trol. *J. Agric. Environ. Ethics*, 12(3): 279-303.

Park, T., P.H. DeBruyn and J.A. Bond, 1958. The relation of X-irradiation to the fecundity and fertility of two species of flour beetles. *Physiol. Zool.*, 31(2):151-170.

Parthasarathy, R. and S.R. Palli, 2009. Molecular analysis of juvenile hormone analogue action in controlling the metamorphosis of the red beetle, *Tribolium castaneum*, *Arch. Insect Biochem. Physiol.*, 70(1): 57-70.

Parthasarathy, R., A. Tan, Z. Sun, Z. Chen, M. Rankin and S.R. Palli, 2009. Juvenile hormone regulation of male accessory gland activity in the red flour beetle, *Tribolium castaneum*. *Mech. Dev.*, 126(7): 563-579.

Parthasarathy, R., R. Farkas and S.R. Palli, 2012. Recent progress in juvenile hormone analogs (JHA) research. In: Advances in Insect Physiology, Insect Growth Disruptors. Dhadialla, T.S. (ed.), Vol. 43, pp. 353-436.

Parthasarathy, R., Z. Sheng, Z. Sun and S.R. Palli, 2010a. Ecdysteroid regulation of ovarian growth and oocyte maturation in the red flour beetle, *Tribolium castaneum. Insect Biochem. Mol. Biol.* 40(6): 429-439.

Parthasarathy, R., Z. Sun, H. Bai and S.R. Palli, 2010b. Juvenile hormone regulation of vitellogenin synthesis in the red flour beetle, *Tribolium castaneum. Insect Biochem. Mol. Biol.*, 40(5): 405-414.

Parween, S. 1996. The effect of Triflumuron on malathion-susceptible (FSS II) and malathion-resistant (CTC 12) strains of *Tribolium castaneum* Herbst. Ph.D. thesis, *University of Newcastle upon Tyne*, UK, pp. 229.

Parween, S. 1997. Effect of triflumuron on the adult midgut of *Tribolium castaneum* Herbst (Coleoptera: Tenebrionidae). *Univ. J. Zool. Rajshahi Univ.*, 16: 11-18.

Pasquier, D. and P.J. Charmillot, 2004. Effectiveness of twelve insecticides applied topically to diapausing larvae of the codling moth, *Cydia pomonella* L. *Pest Manag. Sci.*, 60(3): 305-308.

Pathak, C.S. and S.K. Tiwari, 2010a. Toxicity of neem seed (*Azadirachta indica* A. Juss, Meliaceae) extract against the immature stages of rice-moth, *Corcyra cephalonica* (Staint.) (Lepidoptera: Pyralidae). *J. Appl. Biosci.*, 36 (2): 173-177.

Pathak, C.S. and S.K. Tiwari, 2010b. Toxicological effects of neem *Azadirachta indica* A. Juss leaf powder against the ontogeny of *Corcyra cephalonica* (Staint.) (Lepidoptera: Pyralidae). *J. Biopest.*, 3 (3): 617-621.

Pathak, C.S. and S.K. Tiwari, 2012. Insecticidal action of neem seed (*Azadirachta indica* A. Juss) acetone extract against the life-cycle stages of rice-moth, *Corcyra cephalonica* (Staint.) (Lepidoptera: Pyralidae). *World Appl. Sci. J.*, 8(5): 529-536.

Pavlikova, D. and S. Repas, 1975. Comparative histochemical studies of changes in spermatogenesis and intertubular tissue at male sterility. *Biol. Bratisl.*, 30: 889-895

Pener, M.P. and T.S. Dhadialla, 2012. An Overview of insect growth disruptors; applied aspects. *Adv. Insect Physiol.*, 43: 1-162.

Perveen, F. and T. Miyata, 2000. Effects of sublethal dose of chlorflua-zuron on ovarian development and oogenesis in the com-mon cutworm *Spodoptera litura* (Lepidoptera: Noctuidae). *Ann. Entomol. Soc. Am.*, 93 (5): 1131-1137.

Perveen, F. 2012. Biochemical analyses of action of chlorfluazuron as reproductive inhibitor in *Spodoptera litura*. In: Insecticides-Advances in Integrated Pest Management. *In Tech, Open Access Publisher*, pp. 293-326.

Phillips, D.R. and B.G. Loughton, 1979. The effect of inhibition of RNA and protein synthesis on the development of larvae of *Locusta migratoria. Insect Biochem.* 9(2): 241-245.

Pilo, B., M.V. Asnani and R.V. Shah, 1972. Studies on wound healing and repairs in pigeon liver. III. Histochemical studies on acid and alkaline phosphatase during the process. *J. Anim. Morphol. Physiol.*, 19: 205-212.

Piltz, H. 1977. *Corcyra cephalonica* (Staint.) In: Disease Pests and Weeds tropical crops. Kranz, J., H. Schmutterer and W. Koch (eds.), *Verlag Paul Parey*, Berlin and Hamburg, pp. 439-440.

Pimental, D. 1983. Environmental aspects of pest management. In: Chemistry and World Food Supplies: The New Frontiers, Chemrawn II. L.W. Shemilt (ed.), *Pergamon Press*, Oxford, pp. 185-201.

Pimentel, D., G. Berardi and S. Fast, 1983. Energy efficiency of farming systems: Organic and conventional agriculture. *Agric. Ecosyst. Environ.*, 9:359-372.

Pinto, L.Z., M.M. Bitondi and Z.L. Simoes, 2000. Inhibition of vitellogenin synthesis in *Apis mellifera* workers by a juvenile hormone analogue, pyriproxyfen. *J. Insect Physiol.*, 46(2): 153-160.

Pivnick, K.A. and J.N. McNeil, 1987. Puddling in butterflies: sodium affects reproductive success in *Thymelicus lineola. Physiol. Entomol.*, 12(4): 570-583.

Premkumar, D.R.D., E.P. Jane and S. Mathavan, 1991. Biochemical changes during embryonic development in the aquatic hemipteran bug *Laccotrephes griseus. Insect Biochem.*, 21: 381-388.

Radford, S.V. and D.W. Misch, 1971. The cytological effect of ecdysterone on the midgut cells of the flesh-fly, *Sarcophaga bullata. J. Cell Biol.*, 49(3): 702-711.

Ragonot, E.L. 1885. Revision of the British Species of Phycitidae and Galleridae. *Entomol. Mon. Mag.*, 22: 17-32.

Raikhel, A.S., M.R. Brown and X. Belles, 2005. Hormonal control of reproductive processes. In: Comprehensive Molecular Insect Science. Gilbert L.I., K. Iatrou and S.S. Gill (eds.), Vol. 3, *Elsevier Pergamon*, Oxford, pp. 433-491.

Rajapakse, C.N., R. Meola and J. Readio, 2002. Comparative evaluation of juvenoids for control of cat fleas (Siphonaptera: Pulicidae) in topsoil. *J. Med. Entomol.*, 39(6): 889-894.

Ramamurty, P.S. 1968. Origin and distribution of glycogen during vitellogenesis of the scorpion fly *Panorpa communis. J. Insect Physiol.*, 14: 1325-1330.

Ramaswamy, S.B., S. Shu, Y.I. Park and F. Zeng, 1997. Dynamics of juvenile hormone-mediated gonadotropism in the Lepidoptera. *Arch. Insect Biochem. Physiol.*, 35(4): 539-558.

Reid, B.L., G.W. Bennett and J.W. Yonker, 1990. Influence of fenoxycarb on German cockroach (Dictyoptera: Blattellidae) populations in public housing. *J. Econ. Entomol.*, 83(2): 444-445

Retnakaran, A. 1973. Hormonal induction of supernumerary instars in the spruce budworm, *Choristoneura fumiferana* (Lepidoptera: Tortricidae). *Can. Entomol.*, 105(3): 459-461.

Retnakaran, A., J. Granett and T.J. Ennis, 1985. Insect growth regulators. In: Comprehensive Insect Physiology, Biochemistry and Pharmacology. Kerkut, J.A. and L.I. Gilbert (eds.), Vol. 12, *Pergamon Press*, Oxford, pp. 529-601.

Retnakaran, A. and J.E. Wright, 1987. Control of insect pests with benzoylphenyl ureas. In: Chitin and Benzoylphenyl Ureas. Wright, J.E. and A. Retnakaran (eds.), *Dr. W. Junk Publishers*, Dordrecht, Netherlands, pp. 205-282.

Retnakaran, A. and H. Oberlander, 1993. Control of chitin synthesis in insects. In: Chitin Enzymology, Muzzarelli, R.A.A. (ed.), *European Chitin Society*, pp. 89-99.

Retnakaran, A., K. Hiruma, S.R. Palli and L.M. Riddiford, 1995. Molecular analysis of the mode of action of RH-5992, a lepidopteran-specific, non-steroidal ecdysteroid agonist. *Insect Biochem. Molec. Biol.*, 25(1): 109-117.

Retnakaran, A., A. Macdonald, W.L. Tomkins, C.N. Davis, A.J. Brownright and S.R. Palli, 1997. Ultrastructural effects of a non-steroidal agonist, RH-5992, on the sixth instar larvae of spruce budworm, *Choristoneura fumiferana*. *J. Insect Physiol.*, 43(1): 55-68.

Retnakaran, A., W.L. Tomkins, M.J. Primavera and S.R. Palli, 1999. Feeding behavior of the first-instar *Choristoneura fumiferana* and *Choristoneura pinus pinus* (Lepidoptera: Tortricidae). *Can. Entomol.*, 131(1): 79-84.

Reynolds, S.E. 1987. The cuticle growth and moulting in insects: The essential background to the action of acylurea insecticides. *Pest Sci.*, 20(2): 131-146.

Rharrabe, K., N. Bouayad and F. Sayah, 2009. Effects of ingested 20-hydroxyecdysone on development and midgut epithelial cells of *Plodia interpunctella* (Lepidoptera, Pyralidae). *Pestic. Biochem. Physiol.*, 93(3): 112-119.

Richard, D.S., J.M. Jones, M.R. Barbarito, S. Cerula, J.P. Detweiler, S. J. Fisher, D.M. Brannigan and D.M. Scheswohl, 2001. Vitellogenesis in diapausing and mutant *Drosophila melanogaster*: further evidence for the relative roles of ecdysteroids and juvenile hormones. *J. Insect Physiol.*, 47(8): 905-913.

Richards, O.W. and G.V.B. Herford, 1930. Insects found associated with cacao, spices and dried fruits in London warehouses. *Ann. Appl. Biol.*, 17(2): 367-395.

Riddiford, L.M. and J.W. Truman, 1978. Biochemistry of insect hormones and insect growth regulators. In: Insect Biochemistry. Rockstein, M. (ed.), *Academic Press, Inc.*, New York, pp. 307-357.

Riddiford, L.M. 1994. Cellular and molecular actions of juvenile hormone. I. General considerations and premetamorphic actions. *Adv. Insect Physiol.*, 24: 213-274.

Riddiford, L.M. 1994. A nuclear juvenile hormone-bonding protein from larvae of *Manduca sexta*: A putative receptor for the metamorphic action of juvenile hormone. *Proc. Natl. Acad. Sci. U.S.A.*, 91(13): 6191-6195.

Riddiford, L.M. 1996. Juvenile hormone: the status of its "status quo" action. *Arch. Insect Biochem. Physiol.*, 32(3-4): 271-286,

Riddiford, L.M., K. Hiruma, X. Zhou and C.A. Nelson, 2003. Insights into the molecular basis of the hormonal control of molting and metamorphosis from *Manduca sexta* and *Drosophila melanogaster*. *Insect Biochem. Molec. Biol.*, 33: 1327-1338.

Riddiford, L.M. 2008. Juvenile hormone action: a 2007 perspective. *J. Insect Physiol.* 54: 895-901.

Robertson, J.L. and R.A. Kimball, 1979. Effects of insect growth regulators on the Western spruce budworm (*Choristoneura occidentalis*) (Lepidoptera: Tortricidae): II. Fecundity and fertility reduction following last instar treatments. *Can. Entomol.*, 111: 1369-1380.

Rup, P.J. and P.K. Chopra, 1984. Effect of hydroprene on *Callosobruchus maculatus* (F.) (Coleoptera: Bruchidae). *J. Stored Prod. Res.*, 20(4): 229-232.

Saha, L.M., S. Mandal and D.K. Choudhuri, 1985. Biochemical changes in testis and ovary of *Chrysocoris stolli* Wolf. after the application of juvenoid and ecdysterone. *Rev. Esp. Fisiol.*, 41(2): 249-258.

Saha, L.M., S. Mandal and D.K. Choudhuri, 1986. The effect of juvenile hormone analogue and ecdysterone on the fat body of female *Chrysocoris stollii* Wolf. (Pentatomidae, Heteroptera: Hemiptera). *Zool. Jahrb., Abt. Allg. Zool. Physiol. Tiere*, 90 (1): 85-100.

Samson, P.R., R.J. Parker and E.A. Hall, 1990. Efficacy of the insect growth regulators methoprene, fenoxycarb and diflubenzuron against *Rhyzopertha dominica* (F.) (Coleoptera: Bostrichidae) on maize and paddy rice. *J. Stored Prod. Res.*, 26(4): 215-222.

Samui, T.N., D.C. Deb and S. Chakravorthy, 1981. Note on the effect of hydroprene on the developmental period of the rice moth, *Corcyra cephalonica* Stainton (Lepidoptera: Galleriidae). *Indian J. Agric. Sci.*, 51: 822-824.

Samui, T.N., D.C. Deb and S. Chakravorthy, 1982. Note on the effect of hydroprene on pupal-web formation by *Corcyra cephalonica* Stainton (Lepidoptera: Galleriidae), serious pest of stored grains. *Indian J. Agric. Sci.*, 52(8): 547-548.

Sashindran, N.K., N.J. Sashindran and V.A. Vijayan, 2007. Al-teration in the primary metabolites in three different tis-sues of silkworm, *Bombyx mori* L. under the influence of a juvenoid, R394. *Caspian J. Env. Sci.*, 5(1): 27-33.

Satyanarayana, K., J.Y. Bradfield, G. Bhaskaran and K.H. Dahm, 1994. Stimulation of vitellogenin production by methoprene in prepupae and pupae of *Manduca sexta*. *Arch. Insect Biochem. Physiol.*, 25: 21-37.

Saxena, S.C. and V. Kumar, 1982. Effect of 1-(2, 6-disubstituted benzoyl)-3-phenyl urea compounds on 1st instar larvae of *Trogoderma granarium* (Everts). *Indian J. Entomol.*, 44(1): 49-55.

Saxena, S.C. and D. Kumar, 1991. Effects of two new chitin synthesis inhibitor compounds on different developmental stages of *Trogoderma granarium* Evert. *J. Animal Morph. Physiol.*, 38(1-2): 99-108.

Sazo, L., J.E. Araya and S. Esparza, 2008. Control of San Jose scale nymphs, *Diaspidiotus pernicious* (Comstock), on almond and apple orchards with pyriproxyfen, phenoxycarb, chlorpyrifos, and mineral oil. *Chilean J. Agric. Res.*, 68(3): 284-289.

Schaefer, C.H., W.H. Wilder, F.S. Mulligan III and E.F. Jr. Dupras, 1987. Efficacy of fenoxycarb against mosquitoes (Diptera: Culicidae) and its persistence in the laboratory and field. *J. Econ. Entomol.*, 80 (1): 126-130.

Schmid, A., J.P. Molen, W. Jucker, M. Baggiolini and P. Antonin, 1978. The use of insect growth regulators, analogues of the juvenile hormone, against summer fruit tortrix moth, *Adoxophyes orana* and other pests. *Entomol. Exp. Appl.*,24(1): 65-82.

Schneider, W.C. 1957. Determination of nucleic acids in tissues by pentose analysis. In: Methods in Enzymology. Colowick, S.P. and N.O. Kaplan (eds.), Vol. 3, *Academic Press*, New York, pp. 1154.

Schooley, D.A. and F.C. Baker, 1985. Juvenile hormone biosynthesis. In: Comprehensive Insect Physiology, Biochemistry and Pharmacology. Kerkut, G.A. and L.I. Gilbert (eds.), Vol. 7, *Pergamon*, Oxford, pp. 363-389.

Schuster, L.L. and K.E. Goff, 1995. Dissipation of fenoxycarb in soil under field with and without crop (Squash) in california. *Ciba-Geigy Corporation*. Data Package Report No. 163859-N. DPR No. 51317-090.

Segura, D.F., C. Caceres, M.T. Vera, V. Wornoayporn, A. Islam, P.E.A. Teal, J.L. Cladera, J. Hendrichs and A.S. Robinson, 2009. Enhancing mating performance after juvenile hormone treatment in *Anastrepha fraterculus*: a differential response in males and females acts as a physiological sexing system. *Entomol. Exp. Appl.*, 131(1): 75-84.

Sehnal, F. 1976. Action of juvenoids on different groups of insects. In: The Juvenile Hormones, Gilbert, L.I. (ed.) *Plenum Press, New York*, pp. 301-321.

Sehnal, F. 1983. Juvenile hormone analogs. In: Endocrinology of Insects. Downer R.G.H. and H. Laufer (eds.), *Alan R. Liss. Inc.*, New York, pp. 657- 672.

Sempere, L.F., N.S. Sokol, E.B. Dubrovsky, E.M. Berger and V. Ambros, 2003. Temporal regulation of microRNA expression in *Drosophila melanogaster* mediated by hormonal signals and broad-Complex gene activity. *Dev. Biol.*, 259(1): 9-18.

Seshagir Rao, D. 1954. Notes on rice moth, *Corcyra cephalonica* (Stainton) (Lepidoptera: Pyralidae). *Indian J. Ent.*, 16(2): 95-114.

Sevala, V.L. and K.G. Davey, 1989. Action of juvenile hormone on the follicle cells of *Rhodnius prolixus*: Evidence for a novel regulatory mechanism involving protein kinase C. *Experientia*, 45(4): 355-356.

Shaaya, E. and V. Pisarev, 1986. The lethal effects of three insect juvenile hormone analogues on the developmental stages of *Ephestia cautella* (Wlk.) (Lepidoptera: Phycitidae). *J. Stored Prod. Res.*, 22(3): 125-129.

Shaaya, E., P.D. Shirk, G. Zimowska, S. Plotkin, N.J. Young, H.H. Rees and D.L. Silhacek, 1993. Declining ecdysteroid levels are temporally correlated with the initiation of vitellogenesis dur-ing phareate adult development in Indianmeal moth, *Plodia interpunctella*. *Insect Biochem. Mol. Biol.*, 23(1): 153-158.

Shaffi, S.A. and A.K. Jafri, 1974. The activity of acid and alkaline phosphatase in the egg of some of freshwater teleosts. *Indian J. Fish.*, 21: 286-298.

Shanthy, B., S. Chockalingam and J. Muthukrishnan, 1995. Developmental changes elicited by hydroprene and methoprene in *Tribolium castaneum* Herbst. *Proc. Indian Natl. Sci. Acad.*, B 61(4): 291-298.

Shaurub, E.S.H., Z.A. Ahmed and S.E. El Naggar, 1998. Impacts of pyriproxyfen and extracts of *Schinus terebtnthifolni* Raddi on development, reproduction and reproductive organs on *Spodoptera littoralis* (Boisd.) (Lepidoptera: Noctuidae). *J. Egypt. Ger. Soc. Zool. Ser.*, E 27: 57-82.

Shemshedini, L. and T.G. Wilson, 1990. Resistance to juvenile hormone and an insect growth regulator in *Drosophila* is associated with an altered cytosolic juvenile hormone-binding protein. *Proc. Natl. Acad. Sci. U.S.A.*, 87(6): 2072-2076.

Shirk, P.D., G. Zimowska, D.L. Silhacek and E. Shaaya, 1992. Initiation of vitellogenesis in pharate adult females of the Indianmeal moth, *Plodia interpunctella*. *Arch. Insect Biochem. Physiol.*, 21: 53-63.

Shu, S., Y. Park, S.B. Ramaswamy and A. Srinivasan, 1997. Hemolymph juvenile hormone titers in pupal and adult stages of southwestern corn borer [*Diatraea grandiosella* (Pyralidae)] and relationship with egg development. *J. Insect Physiol.*, 43: 719-726.

Shukla, S. 2010. Studies on the effect certain natural plant products on the chemistry of haemolymph and fat body of the larva of rice-moth, *Corcyra cephalonica* Staint. (Lepidoptera: Pyralidae). Ph.D. Thesis, Deptt. of Zoology, D.D.U. Gorakhpur University, Gorakhpur-273009, U.P., India.

Shukla, S. and S.K. Tiwari, 2011a. Insecticidal role of *Dryopteris filix*-mas ethanolic extract against the developmental stages of rice-moth, *Corcyra cephalonica* Staint. (Lepidoptera: pyralidae). *J. Biopest.*, 4(2): 138-143.

Shukla, S. and S.K. Tiwari, 2011b. Toxicity of *Dryopteris filix*-mas powder against the ontogeny of rice-moth, *Corcyra cephalonica* (Staint.) (Lepidoptera : pyralidae). *Asian J. Exp. Sci.*, 25 (2): 73-79.

Shukla, S. and S.K. Tiwari, 2011c. Toxicological effects of *Dryopteris filix*-mas against the ontogeny of rice-moth, *Corcyra cephalonica* (Staint.). *World Appl. Sci. J.*, 12(1): 16-20.

Shukla, S. and S.K. Tiwari, 2012. The influence of pyrethrum extract on the developmental stages of the rice-moth, *Corcyra cephalonica* Stainton (Lepidoptera: Pyralidae). *Egypt. J. Biol.*, 14: 57-62.

Singh, N. and D.T. Johnson, 2013. Baseline responses of *Alphitobius diaperinus* (Coleoptera: Tenebrionidae) to insect growth regulators. *J. Agric.Urban Entomol.*, 29(1): 35-54.

Singh, P.K. 2010. A decentralized and holistic approach for grain management in India. *Curr. Sci.*, 99(9): 1179-1180.

Singh, S. and K. Kumar, 2011. Diofenolan: a novel insect growth regulator in common citrus butterfly, *Papilio demoleus*. *Phytoparasitica*, 39(3): 205-213.

Slama, K. and C.M.Williams, 1965. Juvenile hormone activity for the bug *Pyrrhocoris apterus*. *Proc. Natl. Acad. Sci. U.S.A.*, 54: 411-414.

Slama, K. and C.M. Williams, 1966. 'Paper factor' as an inhibitor of the embryonic development of the Europian bug, *Pyrrhocoris apterus*. *Nature*, 210: 329-330.

Slama, K., M. Romanuk and F. Sorm, 1974. The chemistry and physiology of juvenoids. In: Insect Hormones and Bioanalogues. *Springer- Verlag*, Berlin, pp. 90-302.

Slama, K. and C.M. Williams, 1996. The juvenile hormone. V. The sensitivity of the bug, *Pyrrhocoris apterus*, to a hormonally active factor in American paper-pulp. *Biol. Bull.*, 130(2): 235-246.

Smet, H., M. Rans and A. de Loof, 1989. Activity of new juvenile hormone analogues on a stored food insect, *Tribolium confusum* (J. du Val) (Coleoptera: Tenebrionidae). *J. Stored Prod. Res.*, 25(3): 165-169.

Smet, H., M. Rans and A. Deloof, 1991. Comparative effectiveness of insect growth regulators with juvenile hormone, anti-juvenile hormone and chitin synthesis inhibiting activity against several stored food insect pests. In: *Proc. Fifth Internat. Working Conference on Stored Product Protection, Bordeaux*, Fleurat-Lessard, F. and P. Ducom (eds.), *Imprimerie du Médoc, Bordeaux*, France, pp. 649-658.

Smith, R.F. and R. van den Bosch, 1967. Integrated Control. In: Pest Control: Biological, Physical and Selected Chemical Method, (eds.) Kilgore W.W. and R.L. Doutt, *Academic Press*, New York, pp: 295-340.

Socha, R. and F. Sehnal, 1972. Inhibition of adult development in *Tenebrio molitor* by insect hormones and antibiotics. *J. Insect Physiol.*, 18(2): 317–337.

Socha, R. and F. Sehnal, 1973. Inhibition of insect development by simultaneous action of prothoracic gland hormone and juvenile hormone. *J. Insect Physiol.*, 19(7): 1449-1453.

Sokal, R.R. and F.J. Rohlf, 1969. Introduction to Biostatistics. *W.H. Freemann and Co.* San Franscisco.

Solomon, M.G. and J.D. Fitzgerald, 1990. Fenoxycarb, a selective insecticide for inclusion in integrated pest management systems for pear in the UK. *J. Hortic. Sci.*, 65(5): 535-539.

Soltani, N. 1984. Effects of ingested diflubenzuron on the longevity and the peritrophic membrane of adult mealworms (*Tenebrio molitor* L.). *Pest. Sci.*, 15(3): 221-225.

Soltani, N., S. Chebira, J.P. Delbecque and J. Delachambre, 1993. Biological activity of flucycloxuron, a novel benzoylphenylurea derivative, on *Tenebrio molitor*: comparison with diflubenzuron and triflumuron. *Experientia*, 49(12): 1088-1091.

Soltani-Mazouni, N. and N. Soltani, 1994. Diflubenzuron affected DNA synthesis in the ovaries of *Tenebrio molitor*. *Invert. Repro. Dev.*, 25(1): 19-21.

Soltani, N. and N. Soltani-Mazouni, 1995. Effect du diflurobenzuron en traitment in vivo et in vitro sur la morphometrie de l'ovarie de *Tenebrio molitor*. *Med. Fac. Landbouww. Univ. Gent.*, 60(3b): 961-968.

Soltani, N. and N. Soltani-Mazouni, 1997. Oogenesis in mealworms: cell density of germarium, thickness of chorion and ecdysteroid production. Effects of regulators. *Med. Fac. Landbouww. Univ. Gent.*, 62(2): 565-571.

Soshkin, D.V. 1991. Effect of diflubenzuron on the mortality of middle age larvae of *Tribolium confusum* and *T. castaneum*. Izv. Timiryazev. S-Kh. Akad., 4:174-176.

Southwood, T.R.E. 1978. Ecological Methods, with Special Reference to the Study of Insect Populations, Second edition, *Chapman and Hall*, London.

Spencer, E.Y. 1974. Organophosphorus Insecticides. In: The Future for Insecticides. Needs and Prospects. Metcalf, R.L. and J.J. Mckelvey, Jr. (Eds.). John Wiley and Sons, New York/London/Sydney/Toronto, pp. 295-307.

Spies, J.R. 1957. Colorimetric procedures for amino acids. In: Methods in Enzymology. Colowick, S.P. and N.O. Kaplan (eds.). *Academic Press*, New York, pp 464-471.

Sridhara, S. and J.V. Bhat, 1963. Alkaline and acid phosphatases of the silkworm, *Bombyx mori* L. *J. Insect Physiol.*, 9: 693 -701.

Srivastav, U.S. and P. Srivastava, 1981. Effect of administration of a juvenoid to the last instar larva of *Earias fabia* Stoll. *Proc. Indian Natl. Sci. Acad.*, B47(6): 799-810.

Staal, G.B. 1972. Biological activity and bioassay of juvenile hormone analogs. In: *Insect* Juvenile Hormones: Chemistry and Action. Menn J.J. and M. Beroza (eds.). *Academic Press Inc.*, London.

Staal, G.B. 1975. Insect growth regulators with juvenile hormone activity. *Ann. Rev. Entomol.* 20: 417-460.

Staal, G.B. 1982. Insect control with growth regulators interfering with the endocrine system. *Entomol. Exp. Appl.*, 31(1): 15-23.

Stainton, H.T. 1866. Description of a New Species of the Family Galleridae. *Entomol. Mon. Mag.*, 2: 172-173.

Steele, J.E. 1961. Occurrence of a hyperglycaemic factor in the corpus cardiacurn of an insect. *Nature*, 192: 680-681.

Stockel, J. and J.P. Edwards, 1981. Susceptibility of *Sitotroga cereallela* (Oliv.) (Lepidoptera: Gelchiidae) to two insect juvenile hormone analogues. *J. Stored Prod. Res.*, 17: 137-141.

Subramanyam, B.H. and D.W. Hagstrum, 1995. Resistance measurement and management. In: Integrated Management of Insects in Stored Products, Subramanyam, Bh. and D.H. Hagstrum (eds.). *Marcel Deckar, Inc.* New York, pp: 331-397.

Sumner, A.T. 1965. The cytology and histochemistry of the digestive gland cells of *Helix*. *Quarl. J. Microsc. Sci.*, 106: 173-192.

Sun, B.B., X.F. Jiang, L. Zhang, D.W. Stanley, L.Z. Luo and W. Lang, 2013. Methoprene influences reproduction and flight capacity in adult of the rice leaf roller, *Cnaphalocrocis Medinalis* (Guenee) (Lepidoptera: Pyralidae). *Arch. Insect Biochem. Physiol.*, 82(1): 1-13.

Sun, X., Q. Song and B. Barrett, 2003. Effect of ecdysone agonists on vitellogenesis and the expression of EcR and USP in codling moth (*Cydia pomonella*). *Arch. Insect Biochem. Physiol.*, 52(3): 115-129.

Tan, N. and K.H. Tan, 1980. Environmental persistency of some juvenile hormone analogues-biological activity against the Mediterranean flour moth, *Ephestia kuehniella* under storage conditions. *Malays. Agric. J.*, 41: 343-350.

Tanaka, S., M.T. Chang, D.L. Denlinger and Y.A.I. Abdel-Aal, 1989. Developmental landmarks and the activity of juvenile hormone and juvenile hormone esterase during the last stadium and pupa of *Lymantria dispar*. *J. Insect Physiol.*, 35(11): 897-905.

Taylor, P. 1980b. Anticholinesterase agents. *In:* The Pharmacological Basis of Therapeutics. Goodman, G.A., *L.S.* Goodman *and A.* Gilman (*eds.),* sixth edition, *New York: Macmillan. pp.* 100-119.

Teal, P.E.A., Y. Gomez-Simuta and A.T. Proveaux, 2000. Mating experience and juvenilehormone enhance sexual signaling and mating in male Caribbean fruit flies. *Proc. Natl. Acad. Sci. U.S.A.*, 97: 3708-3712.

The British Crop Protection Council, 1987. The Pesticide Manual: A World Compendium. Worthing, C.R and R.J. Hance (ed.), Eighth Edition. *The Lavenham Press Limited,* Great Britain.

The British Crop Protection Council, 1991. The Pesticide Manual: A World Compendium. Worthing, C.R. and R.J. Hance (ed.), Ninth Edition, *Unwin Brothers Limited.* Great Britain.

Thede, B.M. 1995a. Aerobic and Aerobic/Anaerobic Metabolism of "A" Label 14C –Fenoxycarb in a Sandy Loam Soil Ciba-Geigy Corporation Data Package Report No. 163859-N. DPR No. 51317-082.

Thede, B.M. 1995b. Aerobic and Aerobic/Anaerobic Metabolism of "A" Label 14C –Fenoxycarb in a Sandy Loam Soil Amendment 1. *Ciba-Geigy Corporation. Data Package Report* No. 163859-N. DPR No. 51317-083.

Thede, B.M. 1995c. Summary of environmental fate studies of fenoxycarb. *Ciba-Geigy. Corporation. Data Package Report* No. 163859-N. DPR No. 51317-077.

Thind, B.B. and J.P. Edwards, 1986. Laboratory evaluation of the juvenile hormone analogue fenoxycarb against some insecticide-susceptible and resistant stored products beetles. *J. Stored Prod. Res.*, 22(4): 235-241.

Thomas, P.J. and P.L. Bhatnagar-Thomas, 1968. Use of a juvenile hormone analogue as insecticide for pests of stored grain. *Nature*, 219: 949.

Thompson, S.N. 2003. Trehalose. The insect 'blood' sugar. *Adv. Insect Physiol.*, 31: 205-285.

Tietze, N.S., P.G. Hester, C.F. Hallmon, M.A. Olson and K.R. Shaffer, 1991. Acute toxicity of mosquitocidal compounds to young mosquito fish, *Gambusia affinis*. *J. Am. Mosq.Control Assoc.*, 7(2): 290-293.

Tiwari, S.K. 1987. Studies on the effect of certain insecticides on the chemistry of haemolymph and fat body of the larva of rice-moth, *Corcyra cephalonica* Stainton (Lepidoptera: Pyralidae). Ph.D. Thesis, Deptt. Zool. Gorakhpur Univ. Gorakhpur-273009, U.P., India.

Tiwari, S.K. and R.S. Bhatt, 1987. Effect of BHC and malathion on the developmental stages of the rice-moth, *Corcyra cephalonica* (Staint.) (Lepidoptera: Pyralidae). *Z. Angew. Zool.*, 74: 83-89.

Tiwari, S.K. and R.S. Bhatt, 1988. BHC and malathion induced changes in glycogen level in haemolymph and fat body of the larva of rice-moth, *Corcyra cephalonica* Staint. (Lepidoptera: Pyralidae). *J. Adv. Zool.*, 9(1): 25-29.

Tiwari, S.K. and R.S. Bhatt, 1989. Effect of BHC and malathion on the total free amino acid level in haemolymph and fat body of the larva of rice-moth, *Corcyra cephalonica* Staint. (Lepidoptera: Pyralidae). *Entomon*, 14(1 and 2): 1330-1337.

Tiwari, S.K. and R.S. Bhatt, 1992. Toxicity of monocrotophos against the ontogeny of rice-moth, *Corcyra cephalonica* (Staint.). *J. Appl. Res.*, 3 (2): 197-198.

Tiwari, S.K. and R.S. Bhatt, 1993. BHC and malathion induced changes in reducing sugar level in haemolymph and fat body of the larva of rice-moth, *Corcyra cephalonica* Staint. (Lepidoptera: Pyralidae). *J. Adv. Zool.*, 14(1):20-24.

Tiwari, S.K. and R.S. Bhatt, 1994a. Toxicity of dimethoate against the ontogeny of rice-moth, *Corcyra cephalonica* (Staint.) (Lepidoptera: Pyralidae). *Bull. Life Sci.*, 4: 101-103.

Tiwari, S.K. and R.S. Bhatt, 1994b. Toxicity of chlordane against the ontogeny of rice-moth, *Corcyra cephalonica* (Staint.) (Lepidoptera: Pyralidae). *Z. Angew. Zool.*, 80: 199-203.

Tiwari, S.K. and R.S. Bhatt, 1994c. Toxicity of methoxychlor against the ontogeny of rice-moth, *Corcyra cephalonica* (Staint.) (Lepidoptera: Pyralidae). *Uttar Pradesh J. Zool.*, 14(1): 52-54.

Tiwari, S.K. and R.S. Bhatt, 1994d. Effect of temephos and tetramethrin on the gonadial biochemistry of rice-moth, *Corcyra cephalonica* (Staint.) (Lepidoptera: Pyralidae). *Z. Angew. Zool.*, 80: 333-344.

Tiwari, S.K. and R.S. Bhatt, 1994e. Effect of chlordane and barthrin on the total protein, total free amino acid and nucleic acid levels in gonadial tissues of rice-moth, *Corcyra cephalonica* (Staint.) (Lepidoptera: Pyralidae). *G. It. Ent.*, 7: 131-135.

Tiwari, S.K. and R.S. Bhatt, 1996. Methoxychlor and dimethoate induced changes in biochemical components of the haemolymph and fat body of the larva of rice-moth, *Corcyra cephalonica* (Staint.) (Lepidoptera: Pyralidae). *Uttar Pradesh J. Zool.*, 16(1): 1-8.

Tiwari, S.K. and R.S. Bhatt, 1999a. Effect of barthrin on the developmental stages of rice-moth, *Corcyra cephalonica* (Staint.) (Lepidoptera: Pyralidae). *J. Adv. Zool.*, 20(2): 103-105.

Tiwari, S.K. and R.S. Bhatt, 1999b. Cypermethrin and fenvalerate induced toxicity against ontogeny of rice-moth, *Corcyra cephalonica. Bull. Life Sci.*, 9: 19-24.

Tiwari, S.K. and R.S. Bhatt, 2000a. Dichlorvos and phosphamidon induced toxicity against ontogeny of rice-moth, *Corcyra cephalonica* (Staint.) (Lepidoptera: Pyralidae). *Uttar Pradesh J. Zool.*, 20(1): 37-40.

Tiwari, S.K. and R.S. Bhatt, 2000b. Effect of cyfluthrin on the developmental stages of rice-moth, *Corcyra cephalonica* (Staint.) (Lepidoptera: Pyralidae). *J. Appl. Zool. Res.*, 11(2/3): 137-139.

Tiwari, S.K. and C.P.M. Tripathi, 2001. Toxicity of temephos against the ontogeny of rice-moth, *Corcyra cephalonica* (Staint.) (Lepidoptera: Pyralidae). *J. Adv. Zool.*, 22(2): 126-128.

Tiwari, S.K. and C.P.M. Tripathi, 2003. Influence of BHC and malathion on the total protein level in haemolymph and fat body of the larva of rice-moth, *Corcyra cephalonica* Staint. (Lepidoptera: Pyralidae). *Uttar Pradesh J. Zool.*, 23(1): 35-37.

Tiwari, S.K. and C.P.M. Tripathi, 2005. BHC and malathion induced changes in nucleic acids level in haemolymph and fat body of the larva of rice-moth, *Corcyra cephalonica* Staint. (Lepidoptera: Pyralidae). *J. Adv. Zool.*, 26(2): 69-72.

Tiwari, S.K. and C.P.M. Tripathi, 2006. Tetramethrin induced toxicity against the ontogeny of rice-moth, *Corcyra cephalonica* (Staint.) (Lepidoptera: Pyralidae). *Uttar Pradesh J. Zool.*, 26(2): 227-229.

Tobe, S.S. and B. Stay, 1985. Structure and regulation of the corpus allatum. *Adv. Insect Physiol.*, 18: 305-432.

Tobe, S.S., R.P. Ruegg, B.A. Stay, F.C. Baker, C.A. Miller and D.A. Schooley, 1985. Juvenile hormone titer and regulation in the cockroach *Diploptera punctata. Experientia*, 41(8): 1028-1034.

Tripathi, P. and S.K. Tiwari, 2013. Effect of methoprene on the reproductive potential of rice moth, *Corcyra cephalonica* (Staint.) (Lepidoptera: Pyralidae). *Am. Int. J. Res. Formal Appl. Natl. Sci.*, 4(1): 33-37.

Tripathi, P. and S.K. Tiwari, 2014. Potential of insect growth regulator in the management of rice moth, *Corcyra cephalonica* Staint. (Lepidoptera: Pyralidae). *Polish J. Entomol.*, 83: 79-97.

Truman, J.W. and L.M. Riddiford, 1974. Hormonal mechanisms underlying insect behavior. *Adv. Insect Physiol.*, 10: 297-352.

Tsumuki, H. and K. Kanehisa, 1984. Phosphatases in the rice stem borer, *Chilo suppressalis* Walker (Lepidoptera, Pyralidae): Some properties and changes of the activities during hibernation. *Cryobiology*, 21(2): 177-182.

Tucker, A.M., J.F. Campbell, F.H. Arthur and K.Y. Zhu, 2014a. Mechanisms for horizontal transfer of methoprene from treated to untreated *Tribolium castaneum* (Herbst). *J. Stored Prod. Res.*, 57: 36-42.

Tucker, A.M., J.F. Campbell, F.H. Arthur and K.Y. Zhu, 2014b. Horizontal transfer of methoprene by *Tribolium castaneum* (Herbst) and *Tribolium confusum* Jacquelin du Val. *J. Stored Prod. Res.*, 57: 73-89.

Tucker, A.M., J.F. Campbell, F.H. Arthur and K.Y. Zhu, 2014c. Efficacy of aerosol applications of methoprene and synergized pyrethrin against Tribolium castaneum adults and eggs. *J. Econ. Entomol.*, 107(3): 1284-1291.

Tunaz, H. and N. Uygun, 2004. Insect growth regulators for insect pest control (Review). *Turkish J. Agric. Forestry*, 28: 377-387.

U.S. Environmental Protection Agency, 1986. Fenoxycarb. Pesticide Fact Sheet Number, 78. USEPA, Washington D.C. pp. 3-37

Valentine, B.J., G.M. Gurr and W.G. Thwaite, 1996. Efficacy of the insect growth regulators tebufenozide and fenoxycarb for lepidopteran pest control in apples, and their compatibility with biological control for integrated pest management. *Aust. J. Exp. Agric.*, 36(4): 501-506.

Van Der Vies, J. 1954. Two methods for the determination of glycogen in liver. *Biochem. J.*, 57(3): 410-416.

Van Pelt-Verkuil, E. 1979. Hormone mediated induction of acid phosphatase activity in the fat body of *Calliphora erythrocephala* prior to metamorphosis. *J. Insect Physiol.*, 25(12): 965-973.

Von Lancker, J.L. and R.L. Holtzer, 1959. The release of acid phosphatase and beta-glucuronidase from cytoplasmic granules in the early course of autolysis. *Am. J. Pathol.*, 35(3): 563-573.

Vorbrodt, A. 1959. The role of phosphatase in intracellular metabolism. *Postepy. Hig. Med. Dosw.*, 13: 200-206.

Wael, L.D., M.D. Greef and O.V. Laere, 1995. Toxicity of pyriproxyfen and fenoxycarb to bumble bee brood using a new method for testing insect growth regulators. *J. Apic. Res.*, 34(1): 3-8.

Walker, P.W. and G.R. Allen, 2010. Mating frequency and reproductive success in an income breeding moth, *Mnesampela privata*. *Entomol. Exp. Appl.*, 136(3): 290-300.

Webb, T.J., S. Shu, S.B. Ramaswamy and S. Dorn, 1999. Influence of juvenile hormone and mating on oogenesis and oviposition in the codling moth, *Cydia pomonella*. *Arch. Insect Biochem. Physiol.*, 41(4):186-200.

Wheeler, D.E. and H.F. Nijhout, 2003. A perspective for understanding the modes of juvenile hormone action as a lipid signaling system. *Bioessays*, 25(10): 994-1001.

Wigglesworth, V.B. 1936. The function of the corpus allatum in the growth and reproduction of *Rhodnius prolixus* (Hemiptera). *Quarl. J. Microsc. Sci.*, 79: 91-121.

Wigglesworth, V.B. 1954. The physiology of insect metamorphosis. Cambridge: *Cambridge University Press*, pp: 1-152.

Wijayaratne, L.K.W. and P.G. Fields, 2010. Effect of methoprene on the heat tolerance and cold tolerance of *Tribolium castaneum* (Herbst) (Coleoptera: Tenebrionidae). *J. Stored Prod. Res.*, 46(3): 166-173.

Wijayaratne, L.K.W., P.G. Fields and F.H. Arthur, 2012a. Effect of methoprene on the progeny production of *Tribolium castaneum* (Coleoptera: Tenebrionidae). *Pest Manag. Sci.*, 68: 217-224.

Wijayaratne, L.K.W., P.G. Fields and F.H. Arthur, 2012b. Residual efficacy of methoprene for control of *Tribolium castaneum* (Coleoptera: Tenebrionidae) larvae at different temperatures on varnished wood, concrete, and wheat. *J. Econ. Entomol.*, 105(2): 718-725.

Wilbur, D.A. and R.B. Mills, 1985. Stored grain insects. In: Fundamentals of Applied Entomology. Pfadt, R.E. (ed.), 4th edition. Macmillan, pp. 552-576.

Williams, P. and T.G. Amos, 1974. Some effects of synthetic juvenile hormones and hormone analogues on *Tribolium castaneum* (Herbst.). *Aust. J. Zool.*, 22 (2): 142-153.

Williams, C.M. 1952. Physiology of insect diapause. IV. The brain and prothoracic glands as an endocrine system in the *Cecropia* silkworm. *Biol. Bull.*, 103(1): 120-138.

Williams, C.M. 1956. The juvenile hormone of insects. *Nature*, 178: 212-213

Williams, C.M. 1961. The Juvenile Hormone. II. Its Role in the Endocrine Control of molting, pupation and, adult development in the *Cecropia* silkworm. *Biol. Bull.*, 121(3): 572-585.

Williams, C.M. 1967. Third generation pesticides. *Sci. Am.*, 217: 13-17.

Williams, D.F., H.L. Collins and O.H. David, 2001. The red imported fire ant (Hymenoptera: Formicidae): an historical perspective of treatment programs and the development of chemical baits for control. *Amer. Entomol.*, 47(3): 146-159.

Williams, D.R., J.H. Chen, M.J. Fisher and H.H. Rees, 1997. Induction of enzymes involved in molting hormone (ecdysteroid) inactivation by ecdysteroids and an agonist, 1, 2-dibenzoyl-1-tert-butylhydrazine (RH-5849). *J. Biol. Chem.*, 272 (13): 8427-8432.

Williams, D.R., M.J. Fisher, G. Smagghe and H.H. Rees, 2002. Species specificity of changes in ecdysteroid metabolism in response to ecdysteroid agonists. *Pestic. Biochem. Physiol.*, 72(2): 91-99.

Willis, J.H. 1974. Morphogenetic action of insect hormones. *Ann. Rev. Entomol.*, 19: 97-115.

Wilson, T.G. and J. Fabian, 1986. A *Drosophila melanogaster* mutant resistant to a chemical analog of juvenile hormone. *Dev. Biol.*, 118(1): 190-201.

Wilson, T.G. 2004. The molecular site of action of juvenile hormone and juvenile hormone insecticides during metamorphosis: how these compounds kill insects. *J. Insect. Physiol.*, 50(2-3): 111-121.

Wing, K.D. and H.E. Aller, 1990. Ecdysteroid agonists as novel insect growth regulators. In: Pesticides and Alternatives: Innovative Chemical and Biological Approaches to Pest Control. Cassida, J.E. (ed.), *Elsevier*, pp. 251-57

WMO (World Meterological Organization), 1994. Scientific Assessment of Ozone Depletion: Proceeding World Meterological Organization Global Ozone Research and Monitoring Project. Report N. 37. WMO, Geneva.

Worthing, C.R. 1987 (Ed.). The Pesticide Manual: A World Compendium. Eighth edition, *The British Crop Procetion council*, Croydon, England.

Wroblewski, V.J., L.G. Harshman, T.N. Hanzlik and B.D. Hammock, 1990. Regulation of juvenile hormone esterase gene expression in the tobacco budworm (Heliothis virescens). *Arch. Biochem. Biophys.*, 278(2): 461-466.

Wyatt, G.R. 1961. The biochemistry of insect hemolymph. *Ann. Rev. Entomol.*, 6(1): 72-102.

Wyatt, G.R. 1967. The biochemistry of sugars and polysaccharides in insects. *Adv. Insect Physiol.*, 4: 287-360.

Wyatt, G.R. 1997. Juvenile hormone in insect reproduction - a paradox ? (Review). *Eur. J. Entomol.*, 94: 323-333. ISSN 1210-5759.

Wyatt, G.R. and K.G. Davey, 1996. Cellular and molecular actions of juvenile hormone. II. Roles of juvenile hormone in adult insects. *Adv. Insect Physiol.*, 26: 1-155.

Xu, Y., F. Fang, Y. Chu, D. Jones and G. Jones, 2002. Activation of transcription through the ligand-binding pocket of the orphan nuclear receptor ultraspiracle. *Eur. J. Biochem.*, 269(24): 6026-6036.

Yamashita, O. and K. Hasegawa, 1985. Embryonic dipause. In: Comprehensive Insect Physiology, Biochemistry and Pharmacology. Kerkut, G.A. and L.I. Gilbert (eds.), Vol.1, *Pergamon Press*, Oxford, pp. 407-434.

Yi, S.X. and T.S. Adams, 2001. Age- and diapause-related acid and alkaline phosphatase activities in the intestine and malpighian tubules of the Colorado potato beetle, *Leptinotarsa decemlineata* (Say). *Arch. Insect Biochem. Physiol.*, 46(3): 152-163.

Zeenath, C. and V.S.K. Nair, 1994. Developmental profile of haemolymph proteins and amino acids in last larval instar of *Spodoptera mauritia* Boisd. following treatment of a juvenile hormone analogue. *Indian J. Exp. Biol.*, 32(4): 271-273.

Zera, A.J. and G. Cisper, 2001. Genetic and diurnal variation in the juvenile hormone titer in a wing-polymorphic cricket: implications for the evolution of life histories and dispersal. *Physiol. Biochem. Zool.*, 74(2): 293-306.

Zhan, S., C. Merlin, J.L. Boore and S.M. Reppert, 2011. The monarch butterfly genome yields insights into long-distance migration. *Cell*, 147(5): 1171-1185.

Zhang, J. and G.R. Wyatt, 1996. Cloning and upstream sequence of a juvenile hormone-regulated gene from the migatory locust. *Gene*, 175(1-2): 193-197.

Zhou, Q., J. Brown, A. Kanarek, J. Rajagopal and D.A. Melton, 2008. *In vivo* reprogramming of adult pancreatic exocrine cells to β-cells. *Nature,* 455: 627-632. doi:10.1038/natureo7314

Zhou, X. and L.M. Riddiford, 2002. Broad specifies pupal development and mediates the 'status quo' action of juvenile hormone on the pupal-adult transformation in *Drosophila* and *Manduca*. *Development,* 129(9): 2259-2269.

Zurflueh, R.C. 1976. Phenylethers as Insect Growth Regulators: laboratory and field experiments. In: Juvenile Hormone. Gilbirt, L.I. (ed.), *Plenum Press,* New York. pp. 61-74.

Index

N

O

P

Q

R

S

T

U

V

W

Y